AA003413

Nanomaterials for Energy Applications

Topical Conference at the 2011 AIChE Annual Meeting

Minneapolis, Minnesota, USA
16-21 October 2011

ISBN: 978-1-61839-578-8

Printed from e-media with permission by:

Curran Associates, Inc.
57 Morehouse Lane
Red Hook, NY 12571

Some format issues inherent in the e-media version may also appear in this print version.

Copyright© (2011) by AIChE
All rights reserved.

Printed by Curran Associates, Inc. (2012)

For permission requests, please contact AIChE
at the address below.

AIChE
3 Park Avenue
New York, NY 10016-5991

Phone: (203) 702-7660
Fax: (203) 775-5177

www.aiche.org

Additional copies of this publication are available from:

Curran Associates, Inc.
57 Morehouse Lane
Red Hook, NY 12571 USA
Phone: 845-758-0400
Fax: 845-758-2634
Email: curran@proceedings.com
Web: www.proceedings.com

TABLE OF CONTENTS

(38a) Designing Nanostructures for High Performance Batteries and Supercapacitors

Yi Cui, Materials Science and Engineering, Stanford University, Stanford, CA

Storing energy electrochemically involves electronic and ionic processes and chemical transformation inside and at the interface of materials. The ability to engineer nanostructures and their interfaces afford the great opportunities for controlling these fundamental processes, which can ultimately lead to high performance devices such as batteries and supercapacitors. Here I will present several examples on designing nanostructures and their interfaces to realize high performance energy storage devices. One example is on designing nanowires and heterostructured nanowires for ultrahigh capacity storage of lithium ions in silicon battery anodes. The challenges of large volume expansion and solid-electrolyte-interphase (SEI) have been overcome. In another example, we have designed nanoporous carbon to encapsulate sulfur for lithium ion battery cathodes, which can addresses many issues related to sulfur. In the third example, paper and textile batteries and supercapacitors have been developed.

See more of this Session: Nanomaterials for Energy Storage I
See more of this Group/Topical: Topical 5: Nanomaterials for Energy Applications

(38b) Nanoscale Organic Hybrid Materials (NOHMs): Opportunities In Electrical Energy Storage

Lynden A. Archer, Chemical and Biomolecular Engineering, Cornell University, Ithaca, NY

This talk focuses on two new classes of nanoparticle organic hybrid materials (NOHMs), which manifest liquid-like ion transport properties at high particle loadings. In one group (Nanoparticle Ionic Materials (NIMs)), electrostatic and steric forces dominate the interparticle potential and appear to be responsible for the system's ability to relax to an equilibrium state. In the other group, only steric forces are present, yet the system is able to reach equilibrium in the absence of a suspending solvent. I will discuss equilibrium structure, rheology, and ionic transport properties of these hybrids. I will also study them for applications as anodes, cathodes, and electrolytes for advanced secondary lithium batteries.

See more of this Session: Nanomaterials for Energy Storage I
See more of this Group/Topical: Topical 5: Nanomaterials for Energy Applications

(38c) Novel Quaternary Ammonium Ionic Liquid Electrolyte for Stable, High Capacity Si Nanowire Based Li-Ion Battery

Vidhya Chakrapani, Florencia Rusli, Michael A. Filler and Paul A. Kohl, School of Chemical & Biomolecular Engineering, Georgia Institute of Technology, Atlanta, GA

*Email: paul.kohl@chbe.gatech.edu

In recent years, there has been interest in silicon nanowires as an anode for Li-ion battery because of its large gravimetric capacity of 4200 mAh/g. 1D nanostructure are essential because they permit radial strain relaxation of mechanical stress which occurs due to the volumetric expansion (>300%) as result of Li insertion. However, commercialization of silicon nanowire-based material has been impeded by number of key issues such as severe capacity fade, limited cycle life and mechanical degradation as result of structural changes. Overcoming these challenges requires optimization of anode structure and choice of better electrolyte than those commonly used. The electrolyte should enable the formation of a stable solid-electrolyte-interface (SEI). Furthermore, the use of organic electrolyte in lithium batteries poses a flammability issue and limits its operating temperature range.

In this work we show a novel quaternary ammonium based ionic liquid as an electrolyte for high capacity and high stability silicon nanowire based anode. Battery testing in these electrolytes showed steady reversible capacity of 2000 mAh/g and columbic efficiency ~97% even after 50 cycles. The efficiency for the first cycle was 91% with very little loss in the discharge capacity indicating good SEI formation in these ionic liquids. Results are discussed in the terms of the nature of this SEI layer and the role of additives that promotes its formation. New strategies were developed to minimize the capacity fade and obtain battery life > 600 cycles.

See more of this Session: Nanomaterials for Energy Storage I
See more of this Group/Topical: Topical 5: Nanomaterials for Energy Applications

Examination of Tropical Cyclogenesis using the High Temporal and Spatial Resolution JRA-25 Dataset

Masato Sugi
Forecast Research Department, Meteorological Research Institute, Japan
Correspondence: msugi@mri-jma.go.jp

INTRODUCTION

Imperfect models alone cannot draw a reliable picture of the real world, while temporally and spatially insufficient observation data alone cannot draw a very detailed picture of the real world. Reanalysis may be regarded as an optimum combination of an imperfect model and insufficient observation data, which can produce the most reliable and detailed dataset to describe the real world.

Regarding the future tropical cyclone (TC) activity, a recent very high resolution GCM experiment indicates a reduction in the TC frequency as a whole, with an increase in the frequency of very intense TCs (Oouchi et al. 2006). On the other hand, in the recent IPCC report, it is summarized as "Based on a range of models, it is likely that future tropical cyclones (typhoons and hurricanes) will become more intense, with larger peak wind speeds and more heavy precipitation associated with ongoing increases of tropical SSTs. There is less confidence in projections of a global decrease in numbers of tropical cyclones." (IPCC, 2007). There is no doubt that for a more reliable projection of the future tropical cyclone activity, it is necessary to better understand the physical processes associated with the genesis and development of tropical cyclones, and to assess the capability of GCMs to simulate these processes.

In the present study, in order to understand the physical processes of tropical cyclogenesis, we examine the formation stage of typhoons using the high temporal and spatial resolution JRA-25 dataset. Because of the high quality of the tropical precipitation in the JRA-25 dataset (Onogi et al. 2006), we can expect tropical disturbances and their environmental circulations are analyzed reasonably well in JRA-25.

DATA

The main data used in the present study is the 6 hourly, 1.25 degree latitude-longitude grid, standard pressure level JRA-25 data. For the JRA-25 precipitation data, 1.125 degree latitude-longitude grid, 6 hour forecast data is used. The CDA file is used to check if the TCR data (tropical cyclone wind profile retrieval data) is used or not in each time. To examine the convective activity around the tropical cyclones, 1 hourly, 0.05 degree latitude-longitude grid GMS satellite black body temperature (TBB) data is used. In addition, data from a very high resolution (horizontal grid size 20km) GCM experiment is used for the purpose of comparison. The period of analysis is the three months from August to October in 2004. During this period 14 typhoons are formed, 8 in August, 3 in September and 3 in October. Formation stages of 12 typhoons out of the 14 typhoons are examined, and three of them, T0416 (CHABA), T0418 (SONGDA) and T0422 (MA-ON), are analyzed in some detail.

DEFINITON OF TC GENESIS

In the present study, we define TC genesis time as the time when TCR data is first used for the TC in JRA-25. TCR data is included in JRA-25 when maximum wind $\geq$ 30 kts in the Neumann's best track data (Hatsushika et al. 2006). On the other hand, the JMA TC genesis time is defined as the time when maximum wind $\geq$ 34 kts, when tropical depression (TD) becomes named typhoon (TS). Because of the difference in the threshold value for TC genesis, the TCR genesis time is usually 6 to 18 hours earlier than the JMA genesis time. In addition, we should note that the maximum winds estimated from satellite images are considerably different in different best track dataset. The maximum wind in JTWC best track data is usually stronger than that in JMA best track data, particularly in the mature stage of TCs. For example, the maximum wind speed of T0416 (CHABA) at mature stage is 110kts in JMA best track data, while it is 155kts in JTWC best track data.

INTENSIFICATION OF VORTEX

We made maps of vorticity at 925hPa every 6 hours from August to October in 2004 over the tropical western North Pacific (Figures not shown). In these maps we can see many weak vortices, some of which develop into typhoons. The central question regarding TC genesis is why some of these vortices develop into TCs but many others do not. We first examine the development processes of the vortices that eventually developed into typhoons. Intensification of vortices during the TC genesis period, from 2 days before the genesis time to 2 days after the genesis time, is shown in Figure 1 for the 12 typhoons formed in August to October in 2004. There was a concern that the inclusion of TCR data might cause a discontinuous intensification of vortex, but the Figure 1 shows that there is no such gaps in the intensity before and after the genesis time. In some case there is a rapid intensification around the genesis time. For example, a rapid intensification can be seen around genesis time of T0419 or T0422, but the intensification continues sometime after the genesis time, and it may be real intensification rather than the artificial one caused by the inclusion of TCR data at genesis time. All the intensification curves of vortices of the 12 TCs are plotted together in Figure 2, with their average. The average curve shows slow intensification before the genesis time and more rapid intensification after the genesis time. This suggests that a self intensification process starts to take place around the genesis time.

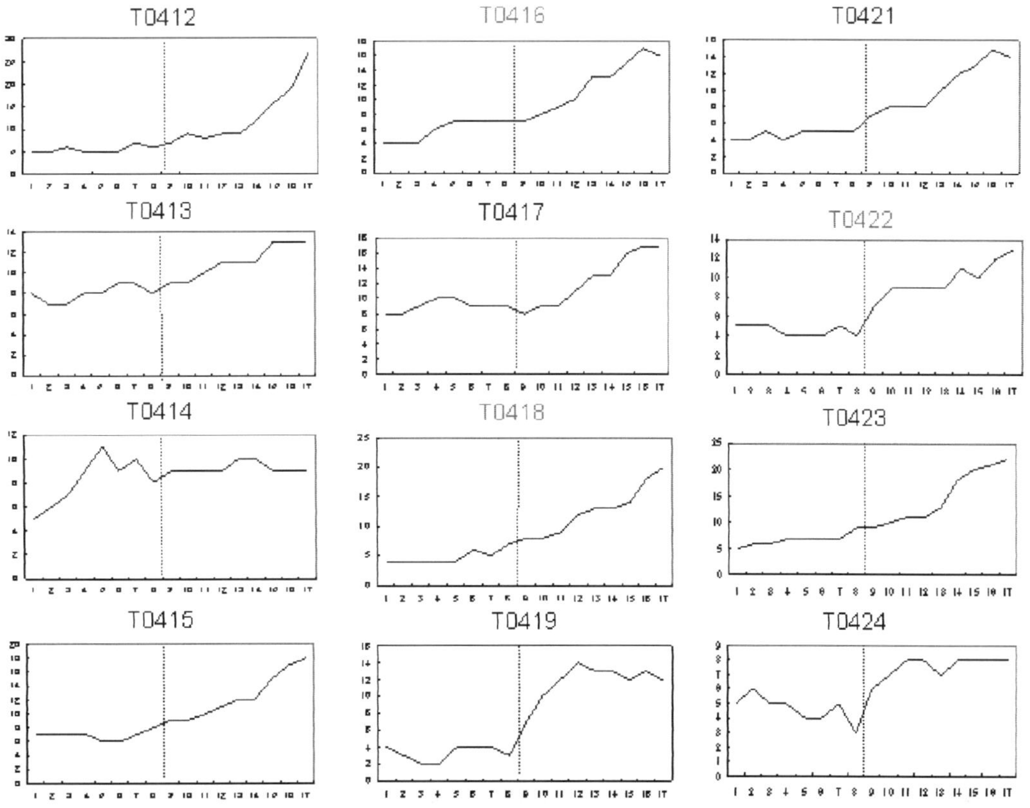

Figure 1 Variation of the intensity of vortex at 925hPa during the 4 day period centered at the genesis time for 12 typhoons formed in August to October of 2004.

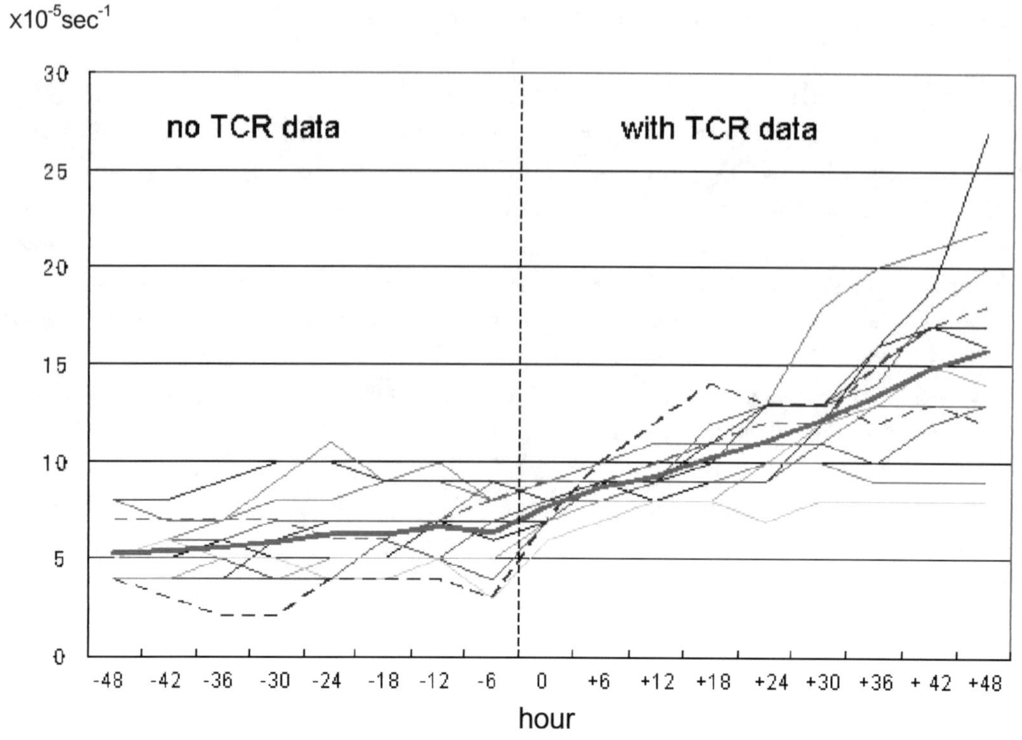

Figure 2 The same as Figure 1 but all curves for 12 typhoons are plotted together, with their average (thick red curve). 0 hour is the genesis time.

CASE STUDIES

We examine genesis processes of three typhoons in some detail. Figure 3 shows the evolution of vorticity, sea level pressure, convective clouds (GMS TBB), precipitation and vorticity generation term $-(\zeta+f)D$ for the TC0416 (CHABA) during the 4 days period centered at the genesis time (18UTC of 18 August 2004). The vertical cross sections of vorticity show the upward development of the vortex from bottom to top, suggesting that the vorticity is generated in the boundary layer and advected upward to free atmosphere by convections. Before the genesis time when the TCR data is first included, a development of vortex together with the surface low pressure area can be seen, indicating that the tropical cyclogenesis process is well captured by the JRA-25 data. Concentrated active convective clouds can be seen near the vortex center after the genesis time, while some scattered active convective clouds can be seen before the genesis time. The role of these scattered convective clouds during the genesis stage seems to be producing a large scale convergence in the boundary layer, leading to an intensification of the vortex through the vorticity generation term. Although the 6-hour forecast precipitations in the JRA-25 are not well concentrated near the vortex center, we can see vorticity generation in the boundary layer near the vortex center. This indicates that the large scale convergence associated with the ensemble of scattered convections around the vortex play an important role in the development of vortex.

The evolutions of the other two typhoons (T0418 and T0422) are almost similar to T0416 (Figures not shown), although there are some minor differences. For both typhoons, we can see upward development of vortex similar to T0416 during the TC genesis period. A development of surface low pressure area and an intensification of vorticity

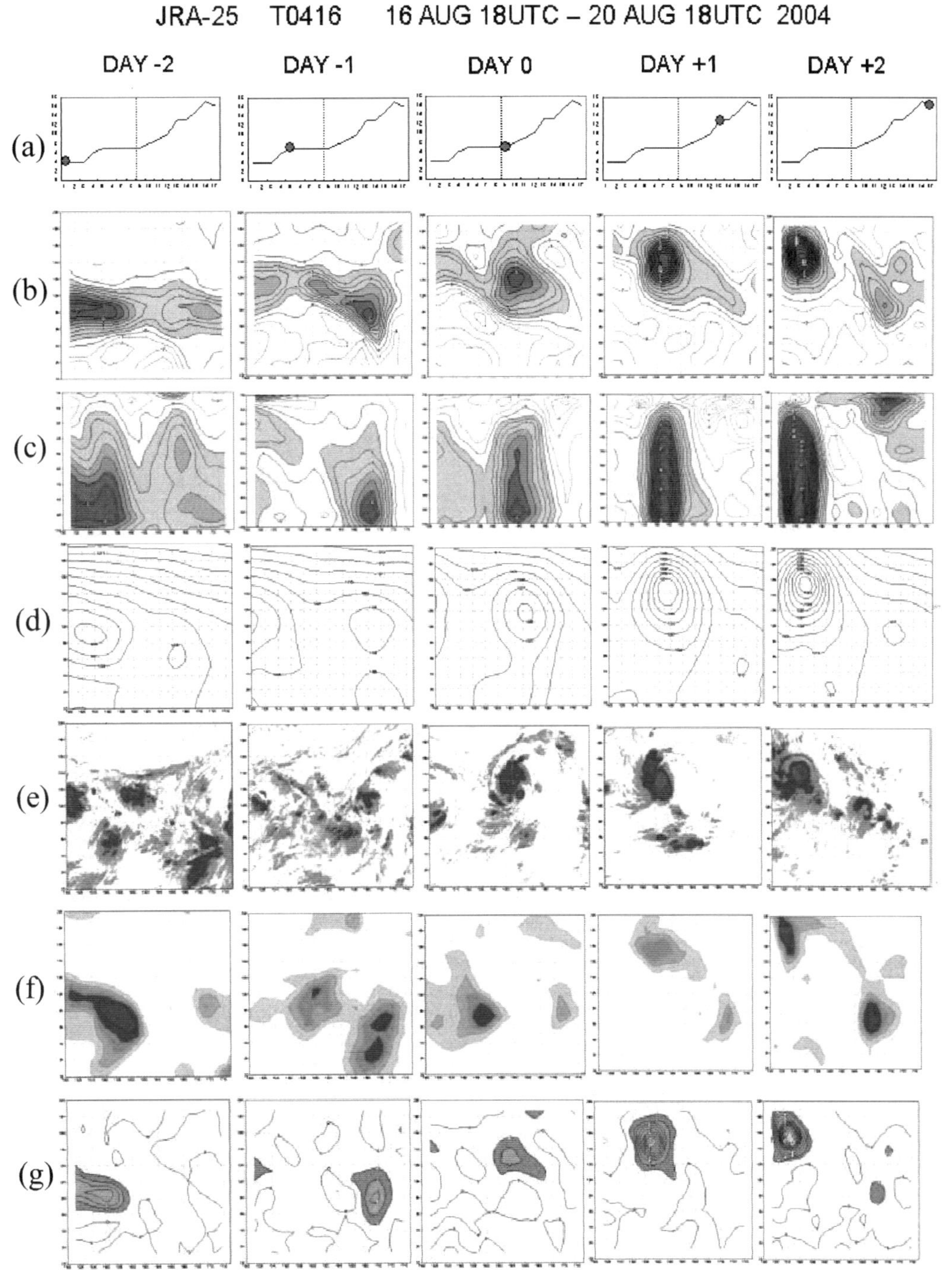

Figure 3 Evolution of T0416 during 4 days centered at the genesis time (18UTC 18 August 2004). (a) Variation of intensity of vortex. (b) Vorticity at 925 hPa. (c) Vertical cross section of vorticity . (d) Sea level pressure. (e) Convective clouds. (f) 6hr forecast precipitation. (g) vorticity generation term ($-(\zeta+f)D$)

generation term around the vortex center can be seen in both cases. We can also see some scattered convective clouds around the vortex center before the genesis time, and concentrated convective clouds near the vortex center as in the T0416 case. Furthermore, a similar evolution of vortex during the TC genesis period is also seen in the very high resolution GCM simulation.2.

CONCLUSIONS

The main conclusions of this study are:

1) High resolution JRA-25 dataset is very useful for TC genesis study. TC genesis process is reasonably well captured by the JRA-25. TCR data makes no significant gaps in the intensity of TC before and after TC genesis.

2) There are many weak vortices ($\leq 5 \times 10^{-5}$ sec^{-1}) in the boundary layer over the tropical western North Pacific. Some of them develop into typhoons but many of them do not develop.

3) It seems that the vorticity generation in the boundary layer through convergence of absolute vorticity ($-(\zeta+f)D$) and its upward advection to free atmosphere by convection are the most important processes for the development of the vortex.. Further quantitative analysis of these processes is an important future work.

4) During the genesis stage of TC, scattered active convective clouds can be seen around the vortex center. The role of these scattered convective clouds during the genesis stage seems to be producing a large scale convergence in the boundary layer, leading to an intensification of the vortex through the vorticity generation term.

5) To assess the capability of GCMs to simulate TC genesis processes, a comparison of GCM simulation with cloud resolving model (CRM) simulation is also an important future work.

ACKNOWLEDGMENT

The author would like to express sincere thanks to the members of JRA-25 project, who helped the author to use the high resolution JRA-25 dataset and CDA files. Thanks are extended to Dr. Mike Fiorino who prepared the high quality TCR data for the JRA-25.

REFERENCES

Hatsushika, H., J. Tsutsui, M. Fiorino, and K. Onogi, 2006: Impact of Wind Profile Retrievals on the Analysis of Tropical Cyclones in the JRA-25 Reanalysis. *J. Meteor. Soc. Japan*, **84**, 891-905.

IPCC 2007: Climate Change 2007: The Physical Science Basis. Cambridge University Press, 996pp

Onogi, K., J. Tsutsui, H. Koide, M. Sakamoto, S. Kobayashi, H. Hatsushika, T. Matsumoto, N. Yamazaki, H. Kamahori, K. Takahashi, S. Kadokura, K. Wada, K. Kato, R. Oyama, T. Ose, N. Mannoji and R. Taira 2007: The JRA-25 Reanalysis. *J. Meteor. Soc. Japan*, **85**, 369-432.

Oouchi, K., J. Yoshimura, H. Yoshimura, R. Mizuta, S. Kusunoki, A. Noda, 2006: Tropical Cyclone Climatology in a Global Warming Climate as Simulated in a 20km-Mesh Global Atmospheric Model: Frequency and Wind Intensity Analysis. *J. Meteor. Soc. Japan*, **84**, 259-276.

(38e) Mechanism of Li Adsorption On Carbon Nanotube-Fullerene Hybrid System: A First-Principles Study

Wonsang Koh[1], Ji Il Choi[2], Kevin Donaher[3], Seung Geol Lee[2] and **Seung Soon Jang**[2], (1)School of Physics, Georgia Institute of Technology, Atlanta, GA, (2)School of Materials Science and Engineering, Georgia Institute of Technology, Atlanta, GA, (3)Department of Applied Physics and Applied Mathematics, Columbia University, New York, NY

The lithium (Li) adsorption mechanism on the metallic (5,5) single wall carbon nanotube (SWCNT)-fullerene (C_{60}) hybrid material system is investigated using first-principles method. It is found that the Li adsorption energy (-2.649 eV) on the CNT-C_{60} hybrid system is lower than that on the peapod system (-1.837 eV) and the bare CNT (-1.720 eV), indicating that the Li adsorption on the CNT-C_{60} hybrid system is more stable than on the peapod or bare CNT system. This is due to the C_{60} of high electron affinity and the charge redistribution after mixing CNT with C_{60}. In order to estimate how efficiently Li can utilize the vast surface area of the hybrid system for increasing energy density, the Li adsorption energy is calculated as a function of the adsorption positions around the CNT-C_{60} hybrid system. It turns out that Li preferably occupies the mid-space between C_{60} and CNT and then wraps up the C_{60} side and subsequently the CNT side. It is also found that the electronic properties of the CNT-C_{60} system, such as band structure, molecular orbital and charge distribution, are influenced by the Li adsorption as a function of the number of Li atoms. From the results, it is expected that the CNT-C_{60} hybrid system has enhanced the charge transport properties in addition to the Li adsorption, compared to both CNT and C_{60}.

See more of this Session: Nanomaterials for Energy Storage I
See more of this Group/Topical: Topical 5: Nanomaterials for Energy Applications

(38f) Vanadium Oxide Nanowire - Carbon Nanotube Binder Free Flexible Electrodes for Supercapacitors

Kenneth J. Balkus Jr. and Sanjaya D. Perera, Chemistry, University of Texas at Dallas, Richardson, TX

Vanadium pentoxide (V_2O_5) layered nanostructures are known to have very stable crystal structures and high faradaic activity. The low electronic conductivity of V_2O_5 greatly limits the application of vanadium oxide as electrode materials and requires combining with conducting materials using binders. It is well known that the organic binders can degrade the overall performance of electrode materials and need carefully controlled compositions. In this study, we develop a simple method for preparing freestanding carbon nanotube (CNT)-V_2O_5 nanowire (VNW) composite paper electrodes without using binders. Coin cell type (CR2032) supercapacitors are assembled for using the nanocomposite paper electrode as the anode and high surface area carbon fiber electrode (Spectracarb 2225) as the cathode. The supercapacitor with CNT-VNW composite paper electrode exhibits a power density of 5.26 kW Kg^{-1} and an energy density of 46.3 Wh Kg^{-1}. Li-VNWs and CNT composite paper electrodes can be fabricated in similar manner and show improved overall performance with a power density of 8.32 kW Kg^{-1} and an energy density of 65.9 Wh Kg^{-1}. The power and energy density values suggest that such flexible hybrid nanocomposite paper electrodes may be useful for high performance electrochemical supercapacitors.

See more of this Session: Nanomaterials for Energy Storage I
See more of this Group/Topical: Topical 5: Nanomaterials for Energy Applications

(39a) Towards Quantum-Dot Solar Cells From Abundant Nontoxic Materials

Ankur Khare[1], Andrew W. Wills[2], Boris D. Chernomordik[1], David J. Norris[3] and **Eray S. Aydil**[1], (1) Chemical Engineering and Materials Science, University of Minnesota, Minneapolis, MN, (2) Department of Chemistry, University of Minnesota Twin Cities, Minneapolis, MN, (3)Optical Materials Engineering Laboratory, ETH Zurich, Zurich, Switzerland

Quantum confinement of electrons and holes in nanometer size crystals (quantum dots or QDs), endows them with properties that may be advantageous for efficient solar-to-electric energy conversion. First, electronic energy levels and optical absorption in QDs can be manipulated by changing their size. This allows the optimization of their optical absorption for maximum overlap with the solar spectrum. Second, the ability to manipulate energy levels through size raises the possibility to make inexpensive multijunction solar cells by judiciously layering different size QDs. Third, it has been suggested that quantum confinement may slow energy dissipative electron and hole relaxation rates such that two new physical processes, multiple exciton generation and hot electron extraction may now compete with relaxation and lead to higher photocurrents or higher photovoltages, respectively. Finally, QDs can be prepared in large quantities as stable colloidal solutions under mild conditions and deposited on surfaces of various planar or nanostructured substrates as thin films through inexpensive high-throughput coating processes to form photovoltaic devices. For these reasons, solar cells based on QDs may have the potential to achieve high power conversion efficiencies at low cost and are promising candidates for third generation photovoltaic devices.

Most quantum dots made to date are based on PbSe and PbS QDs because we know the most about their photophysical properties and synthesis. Lead causes irreversible neurological, reproductive, renal and cardiovascular damage to humans and its use has been banned from many products. Quantum dots based on nontoxic and abundant materials are needed. Copper zinc tin sulfide (Cu_2ZnSnS_4 or CZTS) is emerging as a promising new sustainable semiconductor for photovoltaics. CZTS has a high absorption coefficient in the visible range of the electromagnetic spectrum and an electronic bandgap (~1.5 eV) that is ideal for photovoltaics. We have developed a novel and facile synthesis method for making CZTS nanocrystals and achieved quantum confinement in nanocrystals with diameters less than 3 nm. Our synthesis method is based on decomposition of metal dithiocarbamates in an organic solvent such as dodecene. Presence of oleylamine in the reaction mixture reduces the decomposition temperature of copper, zinc and tin dithiocarbamates to a narrow temperature range (170-220 ^{o}C). Rapid injection of oleylamine and tin diethyldithiocarbamate into a hot mixture of copper and zinc dithiocarbamate in dodecene initiates a burst of nanocrystal nucleation followed by growth. We show that using this approach avoids the formation of binary and ternary sulfide impurities such as zinc sulfide or copper tin sulfide. Moreover, we have been able to control the sizes of CZTS nanocrystals from 2-7 nm by varying the experimental conditions such as the synthesis temperature and time. In this talk we will present our latest results towards making CZTS quantum dot solar cells.

See more of this Session: Nanomaterials for Photovoltaics I
See more of this Group/Topical: Topical 5: Nanomaterials for Energy Applications

(39b) Colloidal Nanomaterials As the Building Blocks for Next-Generation Solar Cells

Jeffrey M. Pietryga, Chemistry Division, Los Alamos National Laboratory, Los Alamos, NM

The use of colloidally synthesized nanomaterials in photovoltaic devices is attractive because of the low-cost and scalability of solution-based fabrication methods. Moreover, the facile control over optoelectronic properties, via the fine control over structure these methods afford, can potentially be exploited in highly efficient next-generation solar cells. Design, synthesis and characterization of novel nanomaterials to enable advanced-concept solar cells are an important part of the ongoing effort within the Center for Advanced Solar Photophysics, a DOE Energy Frontier Research Center. I will discuss a number of specific examples from this work that explore the range of complexities achievable through inexpensive wet-chemical methods. Themes to be examined include the straightforward rendering of established bulk materials into quantum-confined nanocrystal form, band-gap engineering in semiconductor nanoheterostructures, and finally the controlled coupling of plasmonic metal nanoparticles with semiconductor nanocrystals in a diverse series of hybrid superstructures.

See more of this Session: Nanomaterials for Photovoltaics I
See more of this Group/Topical: Topical 5: Nanomaterials for Energy Applications

(39c) Band Gap Tuning with CZTS-Based Nanocrystal-Ink Solar Cells

Grayson Ford[1], Rakesh Agrawal[1] and **Hugh W. Hillhouse**[2], (1)School of Chemical Engineering, Purdue University, West Lafayette, IN, (2)Chemical Engineering, University of Washington, Seattle, WA

$Cu_2ZnSn(S,Se)_4$ is of particular interest due to the achievement of 9.6% efficient laboratory scale devices using a non-vacuum, hydrazine-based deposition process. While hydrazine poses safety challenges from a manufacturing perspective, an alternative non-vacuum deposition approach has achieved 7.2% efficiency via the selenization of Cu_2ZnSnS_4 (CZTS) nanocrystals deposited by roll coating. The resulting $Cu_2ZnSn(S,Se)_4$ (CZTSSe) film has a band gap of approximately 1.1 eV, similar to $CuInSe_2$. $CuInSe_2$ device performance is improved by incorporation of Ga, which widens the band gap. We hypothesized that similar bandgap tuning and improvements to CZTS devices may be possible by substituting a lower atomic number group IV element for some of the Sn. Substitution with germanium is an attractive route, as germanium's crustal abudance of 1.5 mg/kg compares favorably with the 2.3 mg/kg crustal abundance of Sn, the rarest element in CZTS. Here we report a synthesis of $Cu_2Zn(Sn_{1-x}Ge_x)S_4$ (CZTGS) nanocrystals and the performance of $Cu_2Zn(Sn_{1-x}Ge_x)(S_ySe_{1-y})_4$ (CZTGSSe) solar cells fabricated by sintering the nanocrystal films with elemental selenium vapor. We find that the band gap of the CZTGS nanocrystals and the CZTGSSe solar cell can be rationally controlled by adjusting the Ge/(Sn+Ge) ratio. Solar cells fabricated from Cu_2ZnGeS_4 nanocrystal films yielded a power conversion efficiency of 0.51%. However, $Cu_2Zn(Sn_xGe_{1-x})S_4$ nanocrystals with a Ge/(Ge+Sn) ratio 0.7 yielded devices with an efficiency of 6.8% when synthesized to be Cu-poor and Zn-rich. This result opens the possibility of forming Ge gradients to direct minority carriers away from high recombination interfaces and significantly improve the device efficiency of CZTSSe-based solar cells.

See more of this Session: Nanomaterials for Photovoltaics I
See more of this Group/Topical: Topical 5: Nanomaterials for Energy Applications

(39d) Electrostatic Stabilization of Silicon Nanocrystal Colloids for Solution-Processed Photovoltaics

Lance M. Wheeler and Uwe R. Kortshagen, Mechanical Engineering, University of Minnesota, Minneapolis, MN

The tunable optoelectronicproperties of semiconductor nanocrystals (NCs) have garnered significant attention for photovoltaicdevice integration. Additionally,solution processing of semiconductor nanocrystals (NCs) provides a low-cost,scalable technique for the fabrication of third-generation solar cells. Group II-VI and IV-VI compoundsemiconductors have dominated the field of NC photovoltaics, as solution-basedsynthesis processes are well established. The process results in a stericallystabilized colloid from which ligand-capped NCs will naturally self-assemblewhen cast into a film. The ligandscan then be removed or exchanged to form a dense, electronically-coupled NCfilm.

Solution synthesis of Silicon NCshas proven difficult due to the high temperatures needed. Nonthermal flow-through plasmasynthesis offers an effective alternative. Typical plasma processes using silane as the siliconprecursor lead to free-standing, hydrogen-terminated Si NCs that will not,however, stabilize in solution. A hydrosylilation reaction will attach alkylligands to the surface to achieve a stable colloid similar to group II-VI andIV-VI NCs, but post-processing of cast films to remove or exchange the ligandsis problematic.

Here we report nonthermal plasmasynthesis of Si NCs with an alternative silicon precursor, silicontetrachloride. The presence of Clduring synthesis allows for tuning of the surface chemistry of the Si NCs. FTIR analysis reveals Cl present on thesurface, which facilitates a novel colloidal stabilization in aprotic polarsolvents. The Si NCs are readilydispersed and stabilized in aprotic dipolar solvents. The solutions are optically transparent and have remainedstable for over 8 months. Webelieve the electronegativity of the chlorine surface leads to solvation of theNCs by electrostatic interaction with solvent molecules.

Solutions of Cl-terminated Si NCsare spin- or drop-cast into self-assembled films that do not require a post-treatmentto remove ligands. Electronmicroscopy and Scanning Probe Microscopy (SPM) reveal the smooth and continuousfilm morphology needed for photovoltaic device structures. Electrical characterization reveals photoconductiveSi NC films with differential dark conductivities of 1.09×10^{-7}S/cm. This is competitive with nano-and micro-crystalline Si filmsdeposited using high-vacuum techniques such as PECVD.

This work was supported in partsby the National Science Foundation under grant CBET-0756326 and by DOE underthe EFRC Center for Advanced Solar Photophysics.

(39e) Atomic Layer Deposition of Nanoscale Materials for Energy Conversion

Marja N. Mullings, Jonathan R. Bakke, Carl Hgglund and Stacey F. Bent, Chemical Engineering, Stanford University, Stanford, CA

Fora number of technologies, including microelectronics and energy conversiondevices, feature sizes are becoming increasingly small, and the need forprecise control over costly materials continues to grow. In addition to decreasinglength scales, it is important that limited natural resources are carefully used.In fact, many photovoltaic and fuel cell components are comprised of a varietyof materials that may be expensive, toxic and/or rare. Placing such materialsonly where needed to derive the most desirable properties will be essential inreducing the costs of energy conversion devices. Atomic layer deposition (ALD)is uniquely poised to accomplish these goals.

Atomiclayer deposition is a growth technique consisting of self-limiting surfacereactions, which allow for sub-nanometer control of growth. The ability tofinely tune growth allows for careful control over material properties. Furthermore,ALD provides excellent capabilities for depositing a variety of materials onhigh surface area substrates including those that are highly structured.

Thiswork focuses on the ALD and materials characterization of transparentconducting oxides (TCOs) and surface passivation techniques that direct thegrowth of functional materials. Transparent conducting oxides (TCOs) are highlyconductive and transparent materials used in thin film solar cells, serving aselectrical contacts, chemical barriers and antireflection coatings. One of themost common and robust TCO materials, indium-doped tin oxide (ITO), has manyattractive properties; however, alternatives must be found to address thediminishing supplies, increasing demands and subsequently escalating costs ofindium. Replacing indium with zinc to create zinc tin oxide (ZTO) provides acost-effective and earth-abundant alternative. Various material properties ofZTOs are not well understood and there is little control over stoichiometryacross growth systems. At present, it is known that ZTO takes on an unstabletrigonal ilmenite structure ($ZnSO_3$) and stabilizes into the cubic spinel(Zn_2SnO_4) and tin dioxide (SnO_2) under hightemperature conditions.

Bydeveloping and carefully characterizing a ZTO ALD system, these structures willbe better understood and eventually optimized for use as TCOs. Atomic layerdeposition reaction conditions for tin oxide (SnO_x) and zinc oxide(ZnO), which are alternative TCOs, and ZTO will be described, including growthrates and temperature windows in our custom built reactors. Material propertiesof the ZTO films will be characterized with x-ray photoelectron spectroscopy(XPS), scanning Auger electron spectroscopy (AES), spectroscopic ellipsometry,scanning electron microscopy (SEM), x-ray diffractometry (XRD) and four-pointprobe measurements providing insight into surface, structural and electronicproperties.

(39f) Development of a Scalable Bottom-up Nanomanufacturing Platform for Highly Efficient Photovoltaics

Blayne Phillips and **Peng Jiang**, Chemical Engineering, University of Florida, Gainesville, FL

Solar energy is clean, abundant, and renewable, yet it is vastly under-utilized in the current world. Although crystalline silicon solar cells are the premier candidate for renewable and environmentally friendly solar energy conversion, high manufacturing cost has prevented such cells from economically competing with the combustion of fossil fuels as a source of electricity on a large scale. We have developed a scalable bottom-up nanomanufacturing platform for mass-producing highly efficient photovoltaics by exploring effective photon management using surface-plasmon-enabled transparent conducting electrodes and subwavelength antireflective coatings. In this methodology, self-assembled colloidal crystals with remarkably large domain sizes are used as templates to create periodic metallic nanohole arrays and subwavelength moth-eye antireflection coatings. The resulting metallic nanohole arrays exhibit high and tunable optical transmission and high electrical conductivity, promising as transparent conducting electrodes for flexible photovoltaics. The subwavelength-structured moth-eye antireflection coatings show much improved broadband antireflective properties than traditional quarter-wavelength coatings. This novel nanomanufacturing platform combines the simplicity and cost benefits of bottom-up self-assembly with the scalability and compatibility of top-down microfabrication.

See more of this Session: Nanomaterials for Photovoltaics I
See more of this Group/Topical: Topical 5: Nanomaterials for Energy Applications

Materials with Hierarchical Porosity for Energy Storage and Conversion

Andreas Stein, Jane Davidson,
Nicholas Petkovich, Yuqiang Qian, Stephen Rudisill, Luke Venstrom, Anh Vu
University of Minnesota, Minneapolis, MN 55455, U.S.A.

Introduction

Materials containing pores in size ranges from micropores to macropores offer interesting features for a number applications involving energy storage and conversion, such as batteries, fuel cells, hydrogen storage, and sunlight–to–fuel conversion. Depending on the specific application, they can provide large surface areas for reaction, interfacial transport, or dispersion of active sites; they can provide nanostructured features which enhance reactivity, alter materials properties, or shorten diffusion paths; they can act as host materials to stabilize other active components; or, in the case of porous carbons, they can provide electrically conductive phases as well as intercalation sites. However, the higher reactivity brought about by nanostructured features can also lead to decreased stability, particularly in applications where phase changes may alter the structure of the material or where high temperatures are employed. This talk will focus on two redox systems (lithium-ion batteries and sunlight-to-fuel conversion), in which templated porous materials provide a platform for either electrical energy storage or light-to-chemical energy conversion. In the first system the role of pore architecture in carbon-based electrodes will be discussed. In the second system, we investigate the role of porosity in ceria-based materials of interest for solar thermal splitting of water or carbon dioxide to produce hydrogen or carbon monoxide fuels, respectively.

Experimental Section

Three-dimensionally ordered macroporous (3DOM) carbon monoliths and 3DOM carbon with mesoporous walls (3DOM/m C) were synthesized by colloidal crystal templating and nonionic surfactant templating as decribed previously.[1] Tin, tin oxide, sulfur or $LiFePO_4$ were incorporated in the porous carbon materials by solution impregnation, vapor phase impregnation or direct triconstituent syntheses.[2,3] 3DOM CeO_2 and 3DOM $Ce_{0.8}Zr_{0.2}O_2$ powders were synthesized by colloidal crystal templating from $CeCl_3{\cdot}7H_2O$ and $ZrOCl_2{\cdot}8H_2O$ precursors. These samples were compared to a mesoporous CeO_2 sample prepared by thermal decomposition of $Ce(NO_3)_3{\cdot}6H_2O$ in air at 723 K (DS CeO_2) and a commercial polycrystalline ceria sample (Alfa-Aesar, 99.9%). In-situ ceria reduction with hydrogen and oxidation with water or CO_2 were carried out in a fixed bed, vertical tubular flow reactor. Additional details about syntheses, characterization and test reactions are provided in references 1–4.

Results and Discussion

Improved rate capabilities for lithiation/delithiation are observed for hierarchically porous carbon electrodes. In composites with tin or tin oxide for anodes of lithium ion batteries, these structures maintain electrical contact between tin-based particles, even when those particles undergo significant volume changes during cycling, and hence the composite anode maintains good capacities over multiple cycles (Figure 1). Composites of hierarchically structured carbon with poorly conducting but otherwise desirable electrode materials (like sulfur or $LiFePO_4$) can be used to overcome limitations in electrical conductivity of those materials, increasing the choice of useful electrode materials.

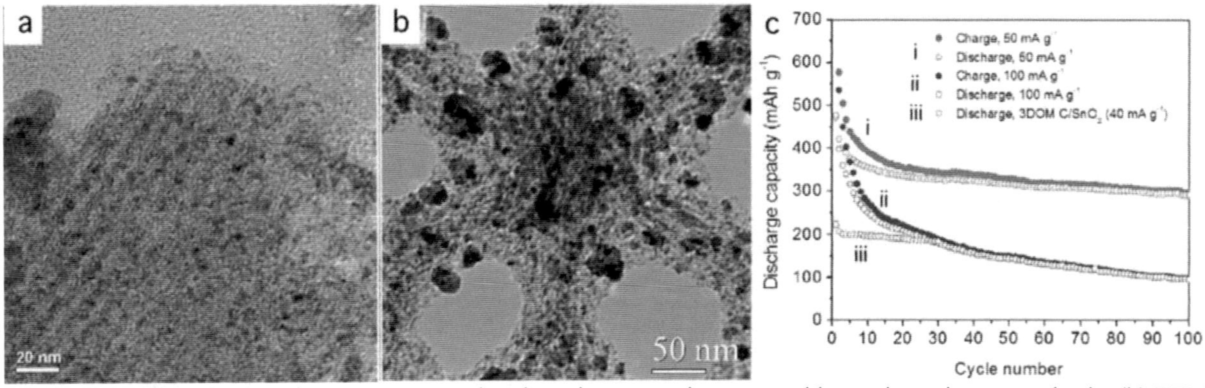

Figure 1. (a) TEM image of a mesostructured carbon-tin composite prepared by a triconstituent synthesis. (b) TEM image of a carbon-tin oxide composite structure with hierarchical porosity. (c) Lithiation charge and discharge curves of 3DOM/m C/SnO_2 (i), 3DOM C coated with SnO_2 (ii) and 3DOM carbon (iii) as function of cycle number.[2]

In reactions targeting the production of chemical fuels (H_2, CO) from water and carbon dioxide, the heterogeneous conversion processes benefit from added surface area in 3DOM CeO_2 samples that facilitate the oxidation cyle. While overall productivity does not depend strongly on sample morphology (given sufficient reaction time), the addition of porosity improves peak fuel production rates and reactivities (Figure 2). However, a mesoporous ceria sample displaying an even higher surface area (DS CeO_2) quickly lost activity as the material sintered during the reaction. The loss in surface area due to sintering of 3DOM CeO_2 was less and could be improved further by doping of ceria with zirconia, which stabilized the porous structures at the high reaction temperatures. The interplay between composition and morphology of these materials, thermal stability, and conversion efficiencies will be discussed.

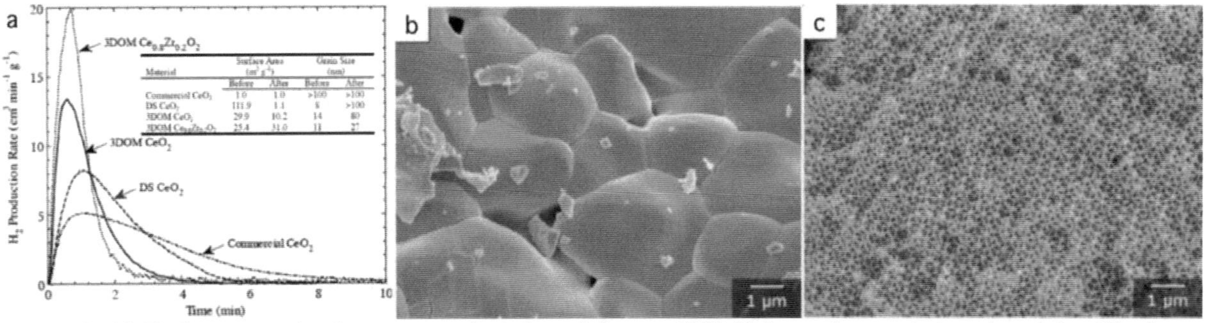

Figure 2. (a) Hydrogen production rate as a function of time at 1073 K for various ceria-based samples. The inset tabulates specific surface areas and grain sizes of the various ceria-based samples before redox cycling and after redox cycling at 1073 K. The SEM images show (b) a chemically cycled commercial CeO_2 sample and (c) a cycled 3DOM $Ce_{0.8}Zr0._{2}O_2$ sample.[4]

References

1. Wang, Z.; Stein, A. *Chem. Mater.* **2008**, *20*, 1029-1040
2. Wang, Z.; Fierke, M. A.; Stein, A. *J. Electrochem. Soc.*, **2008**, *155*, A658-A663.
3. Vu, A.; Stein, A. *Chem. Mater.* **2011**, *23*, 3237-3245.
4. Venstrom, L.; Petkovich, N.; Rudisill, S.; Stein, A.; Davidson, J. H. *J. Sol. Energ.-T. ASME* **2011**, submitted.

(106b) Investigation of Charge Storage Mechanism of Nanostructured Metal Nitrides and Carbides Electrodes for Electrochemical Capacitors

Priyanka Pande, Paul G Rasmussen and Levi T. Thompson, Department of Chemical Engineering, University of Michigan, Ann Arbor, MI

Electrochemical capacitors (ECs) offer aunique combination of high power and energy densities filling the gap betweenconventional capacitors and batteries. These devices store energy via doublelayer (i.e. electrostatic mechanism) and/or pseudo-capacitance mechanisms thatinvolve charge transfer processes [1]. Early transition metal nitrides andcarbides are promising candidates for use as electrode materials due to theirhigh electronic conductivities, surface areas (up to 200 m^2/g) andelectrochemical stabilities [2,3]. For example, vanadium nitride has beenreported to have the highest capacitance up to 1340 Fg^{-1} [4].However the charge storage mechanism for these materials is not wellunderstood. In this paper we will report results from ion isolation experiments,impedance spectroscopy, charge-discharge and chronotentiometry and suggest chargestorage mechanisms that reconcile the results.

Nanostructured V, Mo, W nitrides andcarbides were synthesized via temperature-programmed-reaction of their oxideprecursors with anhydrous NH_3 or 15% CH_4/H_2followed by passivation in 1% O_2/He at room temperature to form aoxygen-rich passivation layer preventing bulk oxidation on exposure to air[2,3]. Physical characterization was performed using BET surface area analysis,scanning electron microscopy and X-ray diffraction. The stability of these materials in aqueous KOH and $H_2SO_4$was determined by cyclic voltammetry. The species contributing to chargestorage were identified by isolation of the electrolyte ions and pairing themwith inactive counter ions. H^+ and SO_4^{2-}, wereisolated as $H^+BF_4^-$ (tetrafluoroboric acid) and$[(C_2H_5)_4N^+]_2\,SO_4^{2-}$ (tetraethylammoniumsulfate, $(TEA)_2SO_4$), while K^+ and OH^-ions were isolated as $K^+(CF_3SO_3)^-$(potassium triflate, K-triflate) and $(C_2H_5)_4N^+OH^-$(tetraethylammonium hydroxide, TEA-OH). Cyclic voltammetry experiments werecarried out in each of these electrolytes. In order to establish the chargetransfer reaction chronopotentiometry experiments were carried out in varyingelectrolyte concentrations.

The redox behaviorconfirmed that most of the charge storage was via the pseudocapacitancemechanism. The ion isolation experiments indicated that OH^- and H^+were the primary species contributing to charge storage in VN in KOH (figure 1)and Mo_2N in H_2SO_4 systems, respectively. Thechronopotentiometry results (figure 2) suggest that the charge storage reactionfor VN was

$$VN_xO_y - 2OH^- \leftrightarrow [VN_xO_y(OH)_2]^- - e^-$$

Results from impedancespectroscopy and three-electrode charge-discharge experiments will also bepresented

REFERENCES

[1] SimonP, Burke A, The ElectrochemicalSociety: Interface, Spring(2008) 38.

[2] Cladridge J B, York A P E,Brungs A J, Green Malcolm L H, Chem. Mater.12(2000) 132.

[3] WixomM R, Tarnowski D J, Parker J M, Lee J Q, Chen P -L, Song I, Thompson L T, Mat.Res. Soc. Symp. Proc. 496 (1998) 643.

[4] Choi D, Kumta P N, Electrochem. Solid-StateLett. 8 8 (2005) A418.

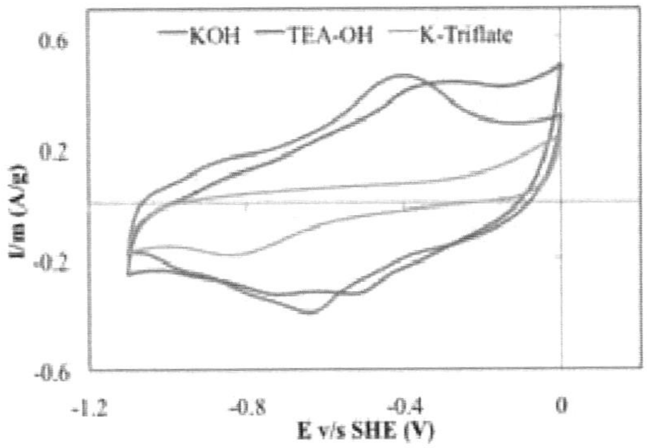

Figure 1: Cyclic voltammogram for VN in various electrolyte solutions at a scan rate of2mVs^{-1}.

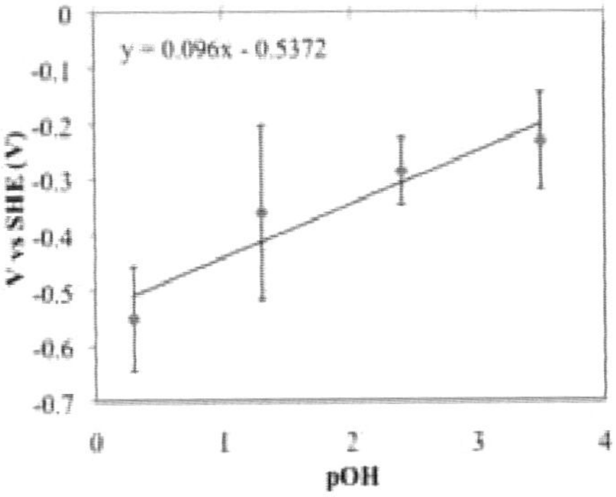

Figure 2: pOH *vs*Voltage for VN in KOH. Slope of 0.096 indicates that there are 2 OH^- ions reacting per electron transferred.

See more of this Session: Nanomaterials for Energy Storage II
See more of this Group/Topical: Topical 5: Nanomaterials for Energy Applications

(106c) High-Performance Electrochemical Capacitors Based On Hybrid Graphene/Metal Oxide Nanostructured Textiles

Guihua Yu[1], Liangbing Hu[2], Yi Cui[2] and Zhenan Bao[1], (1)Chemical Engineering, Stanford University, Stanford, CA, (2)Materials Science and Engineering, Stanford University, Stanford, CA

Large scale energy storage system with low cost, high power and long cycle life is crucial for addressing the energy problem when connected with renewable energy production. To realize large-scale applications of the energy storage devices, there remain several key issues including the development of low-cost, high-performance materials that are environmentally friendly and compatible with low-temperature and large-scale processing. In this talk, we will present a novel structure based on graphene, transition metal oxide and textiles where solution-exfoliated graphene nanosheets conformably coated on highly porous textiles function as three-dimensional conductors and electrodeposited metal oxide nanostructures function as active electrode materials. Electrochemical capacitors based on hybrid graphene/metal oxide based textiles in aqueous electrolytes exhibit very promising characteristics with a power density of ~110 kW/kg, and excellent cycling performance of >95% capacitance retention over 5000 cycles, which are substantially better than those previously reported for this oxide-based system. Such low-cost, high-performance energy textiles based on all-solution-processed graphene/metal oxide hierarchical nanostructures offer great promise in grid-scale energy storage device applications.

See more of this Session: Nanomaterials for Energy Storage II
See more of this Group/Topical: Topical 5: Nanomaterials for Energy Applications

(106d) MnO2 Doped Graphene As Electrodes in Supercapacitor

Lixin Wang[1], Kapila Wadumesthrige[1], Steven O. Salley[2] and K. Y. Simon Ng[3], (1)Chemical Engineering, Wayne State University, Detroit, MI, (2)Chemical Engineering and Material Science, Wayne State University, Detroit, MI, (3)Wayne State University, Detroit, MI

Abstract:

Graphene doped carbon nanotubes for high-energy supercapacitors

A supercapacitor is an electrochemical capacitor with a relatively high energy density and is characterized by fast discharge rates and low maintenance. These characteristics have already been explored in various applications such as consumer electronics, wind energy system, smart grid applications and electric vehicles. The primary purpose behind the burgeoning effort in supercapacitor research is to achieve greater energy densities at a lower cost. Vertical carbon nanotube (CNT) is a promising electrode material for supercapacitor production with a superior conductivity, large voltage window of stability, high surface area, steady mechanical strength and low cost. Capacitance values up to 135 F/g were reported with multi-walled carbon nanotubes. However, the relatively small inherent capacitance of CNTs confines their use to supercapcitor applications. In this study, using different methods, graphite will be doped into CNT walls to improve its capacitance. The doped CNTs will then be used in supercapacitor electrode applications and their performance tested in order to find a suitable method of improving overall capacitance. 200 F/g capacitance is expected after the optimization of doping materials and doping methods.

See more of this Session: Nanomaterials for Energy Storage II
See more of this Group/Topical: Topical 5: Nanomaterials for Energy Applications

(106e) Thermopower Wave Dynamics: Nanoscale Alternating Current Sources and Geometry Effects

Joel T. Abrahamson[1], Wonjoon Choi[2], Sayalee G. Mahajan[1], Nicole S. Schonenbach[1], Jungsik Park[1], Michael P. Walsh[1], Jae-Hee Han[1], Kourosh Kalantar-Zadeh[3] and Michael S. Strano[1], (1)Department of Chemical Engineering, Massachusetts Institute of Technology, Cambridge, MA, (2)Mechanical Engineering, Massachusetts Institute of Technology, Cambridge, MA, (3)School of Electrical and Computer Engineering, RMIT University, Melbourne, Australia

The nonlinear coupling between exothermic chemical reactions and a nanostructure with large heat conduction results in a self-propagating thermal wave guided along the nano-conduit in the direction of its greatest thermal diffusivity [1,2]. The resulting reaction wave pushes charge carriers along, creating a concomitant thermopower wave and electrical current in the same direction. For carbon nanotubes, examples of one-dimensional conduits, we have measured high power density (> 7 kW/kg) from such waves. However, two-dimensional conduits such as graphene should also provide new avenues for control of wave velocity and temperature due to additional degrees of control from geometry and boundaries. We have developed the theory and mathematics of such waves for one- and two-dimensional conduits.

For one-dimensional conduits, our models predict that certain values of the chemical reaction kinetics and thermal parameters produce oscillating reaction wave velocities with well-defined frequency and amplitude [3]. We demonstrate such oscillations experimentally using a cyclotrimethylene-trinitramine/ multiwalled carbon nanotube system, which produces frequencies in the range of 400 to 5000 Hz. The propagation velocity oscillations and the frequency dispersion are well described by our theoretical calculations and are linked to oscillations in the voltage generated by the reaction. These thermopower oscillations may enable new types of nanoscale power and signal processing sources.

[1] W. Choi, S. Hong, J. T. Abrahamson, *et al. Nature Materials*, 9, (2010), 423.

[2] W. Choi, J. T. Abrahamson, J. M. Strano, M. S. Strano. *Materials Today,* 13, (2010), 22.

[3] J. T. Abrahamson, *et al. ACS Nano* 5, (2011), 367.

See more of this Session: Nanomaterials for Energy Storage II
See more of this Group/Topical: Topical 5: Nanomaterials for Energy Applications

(107a) Electronic Transport In Solid of Colloidal Quantum Dots

Philippe Guyot-Sionnest, James Franck Institute, University of Chicago, Chicago, IL

Colloidal quantum dots can be made of a wide variety of materials, covering the IR to the UV spectral range. They can be processed as liquid solutions, dried, stamped, printed, on many different substrates. One of our original contribution has been to achieve conductivity in solids of rather monodispersed colloidal quantum dots. This was made possible by controlling the interaction between the dots and the Fermi level, leading to reversibly tuning the electronic transport properties over many orders of magnitude. Our particular interests are to understand the basic mechanisms and transport regime associated with the nature of the materials. Starting with well monodispersed materials allows to control accurately the carrier density by temperature/charging/doping, to monitor the effet on the occupation of the electronic levels, and the effect of that occupation on the radiative and nonradiative properties of the materials. These studies provide the scientific basis for future optoelectronic application of solids of colloidal quantum dots.

See more of this Session: Nanomaterials for Photovoltaics II
See more of this Group/Topical: Topical 5: Nanomaterials for Energy Applications

(107b) Role of Quantized and Mid-Gap States In "Dark" Charge Transport and Photoconductivity In Semiconductor Nanocrystal Films

Prashant Nagpal, Physical Chemistry and Applied Spectroscopy, Los Alamos National Labratory, Los Alamos, NM

Colloidal semiconductor nanocrystals (NCs) have attracted significant interest for applications in solution-processable devices such as light-emitting diodes and solar cells. However, poor understanding of charge transport in NC assemblies, specifically the relation between electrical conductance in dark and under light illumination, hinders their technological applicability. Here, we simultaneously address the issues of "dark" transport and photoconductivity in NC films by incorporating them into optical field-effect transistors (OFETs), in which the channel conductance is controlled by both gate voltage and incident radiation. Spectrally resolved photoresponses of OFETs reveal a weakly conductive mid-gap band (MGB) which is responsible for charge transport in dark. The mechanism for conductance, however, changes under illumination when it becomes dominated by band-edge quantized states. In this case, the MGB still plays an important role as its occupancy (tuned by the gate voltage) controls the dynamics of band-edge charges. Our study has broad implications for NC-based electronics and optoelectronics, and specifically, suggests that design guidelines for NC devices might be different depending on whether they are intended for operation in dark (diodes and transistors) or under illumination (photodetectors and solar cells).

See more of this Session: Nanomaterials for Photovoltaics II
See more of this Group/Topical: Topical 5: Nanomaterials for Energy Applications

(107c) Molecular Simulation of Ligand-Capped Lead Selenide Nanocrystals

Ananth P. Kaushik, Alex Tait and Paulette Clancy, Chemical and Biomolecular Engineering, Cornell University, Ithaca, NY

Nanocrystalline solids have become the subject of intense study due to their unique optical properties and their capacity to form self-assembled superlattices. These properties make them suitable for use in a variety of applications such as solar cells, light emitting devices, transistors, etc. The self-assembly of these nanoparticles is governed by interactions at the molecular level, and hence, understanding the nature of these interactions could be instrumental in achieving precise control over the self-assembly process. The nanocrystals are "capped" by organic molecules which are believed to drive the self-assembly and stabilize the final superlattice and which are responsible for most of the interparticle forces. Despite the importance of the ligand-ligand interactions, there is very little fundamental understanding of these hybrid systems. In our work, we use atomically and molecularly explicit Molecular Dynamics simulations to understand the interactions between the ligands on the surface of the nanocrystals. Specifically, we observe the energetic interactions of experimentally sized nanocrystals capped with ligands in various lattices including face-centered cubic (fcc) and body-centered cubic (bcc) structures in collaboration with the Hanrath group at Cornell. We also undertake the study of key controlling parameters of self-assembly including nanocrystal diameter, ligand grafting density, ligand length, nanocrystal orientation and size of nanocrystal facets, etc.

These results form the first report of computer simulations of experimentally relevant sized ligand-capped nanocrystals (3-6 nm). All prior simulation work in the literature has focused on nanocrystal diameters of less than 3 nm. The importance of nanocrystal size was clearly uncovered in our studies, which showed that the high ratio of ligand length to crystal facet size for 3 nm particles facilitates the wrapping of ligands around the nanocrystal cores. This, in turn, produces an effectively spherical interaction among nanocrystals that is orientation-independent. For larger nanocrystals (4 and 6 nm), the lower ligand-to-diameter ratio allows many ligands to experience more face than edge interactions and tend to stick out from the facet rather than wrapping around the crystal. Our simulations have also shown that these interactions are dependent on the grafting density of ligands and on the density of ligands on various facets of the nanocrystals.

We are also investigating the orientational alignment of nanocrystals in superlattices and the effect of local asymmetries in ligand density on the selection of preferred morphologies. The alignment of nanocrystals in the superlattice is primarily due to the orientations of the capping ligands. And the anisotropy of ligand coverage is responsible for orientational alignment of the nanocrystals, which, in turn, drives superlattice symmetry.

The choice of solvents on ligand morphologies also plays a role; the solvent appears to be responsible for superlattice stability as well superlattice symmetry. The solvent may also be responsible for different ligand morphologies on the nanocrystal surface. This kind of study provides key insights into molecular-scale information about ligand interactions which can be leveraged into more coarse-grained mesoscale simulations of nucleation and growth phenomena which may to be used to predict and guide future experimental studies.

See more of this Session: Nanomaterials for Photovoltaics II
See more of this Group/Topical: Topical 5: Nanomaterials for Energy Applications

(107d) Analyzing Phase Transformation In the Growth of CIGSe Films From Nanocrystals

Bryce C. Walker, Erik Sheets and Rakesh Agrawal, School of Chemical Engineering, Purdue University, West Lafayette, IN

Copper Indium diSelenide (CISe) and its derivative Copper Indium Gallium diSelenide (CIGSe) have gained increasing amounts of attention as materials for photovoltaic energy conversion. Recently CIGSe devices have proven to be able to produce power generation efficiencies exceeding 20% on the laboratory scale [1].

To increase the viability of CIGSe based devices there has recently been a push to decrease costs associated with the production of CISe/CIGSe solar cells through the utilization of Cu, In and Ga containing nanocrystals. Incorporation of nanomaterial based formation allows the utilization of solution based processing resulting in more scalable processes. One approach demonstrated by Guo et al investigates the incorporation of selenium into sulfide based CIGS (copper indium gallium disulfide) nanoparticles for grain growth to dense crystal structures with power generation efficiencies exceeding 10% [2].

Building off the work of Guo et al, the growth mechanisms of films from their respective Cu, In, and Ga containing nanoparticles are developed and shown. Work incorporates both in-situ and ex-situ experiments with the grain growth observed, and the applications of knowledge to other chalcopyrite based solar cells are discussed.

[1] M.A. Green, K. Emery, Y. Hishikawa, and W. Warta, "Solar cell efficiency tables (version 37)," *Progress in Photovoltaics: Research and Applications*, vol. 19, Jan. 2011, pp. 84-92.

[2] Q. Guo, G.M. Ford, H.W. Hillhouse, and R. Agrawal, "Sulfide nanocrystal inks for dense Cu (In1-xGa(x))(S1-ySe(y))2 absorber films and their photovoltaic performance.," *Nano letters*, vol. 9, Aug. 2009, pp. 3060-5.

See more of this Session: Nanomaterials for Photovoltaics II
See more of this Group/Topical: Topical 5: Nanomaterials for Energy Applications

(107e) Multilayer Assembly of Nanowire Arrays for Dye-Sensitized Solar Cells

Chengkun Xu, Jiamin Wu, **Umang V. Desai** and Di Gao, Chemical and Petroleum Engineering, University of Pittsburgh, Pittsburgh, PA

Vertically ordered nanostructures synthesized directly on transparent conducting oxide have shown great promise to overcome the limitations of current dye-sensitized solar cells (DSCs) based on random networks of nanoparticles. Synthesis of such structures with a high internal surface area, however, has been challenging. Here we demonstrate a convenient process for synthesizing multilayer assemblies of ZnO nanowire arrays and using the assemblies for fabrication of DSCs. The assembled multilayer ZnO nanowire arrays possess an internal surface area that is more than five times larger than what one can possibly obtain with single-layer nanowire arrays. DSCs fabricated using such multilayer arrays yield a power conversion efficiency of 7%, which is comparable to TiO2 nanoparticle-based DSCs. The ordered structure with a high internal surface area opens up opportunities for further improvement of DSCs by using solid-state electrolytes and kinetically fast redox couples.

See more of this Session: Nanomaterials for Photovoltaics II
See more of this Group/Topical: Topical 5: Nanomaterials for Energy Applications

(107f) Inverse Opal Topology of Titania Electrodes for Photovoltaics

Sonia S. Mathew, Chemical Engineering, City College of New York, New York, NY and Ilona Kretzschmar, Chemical Engineering, The City College of New York, The City University of New York, New York, NY

Dye-sensitized solar cells use porous titania photoanodes to convert solar energy to usable work. Traditionally, the titania electrode is a randomly connected network of titania nanoparticles. This configuration, while it maximizes the surface area for dye-adhesion, which in turn maximizes the amount of energy harnessed, is also prone to excessive recombination. Our approach is to improve the efficiency of the DSSC is by minimizing these recombination events by introducing order to the titania electrode. Our electrode is made of titania inverse opals with pore sizes on the order of 100s nm, compared to <20 nm in the current setup. This configuration avoids incomplete filling of the pores, which hampers the electron transfer across the interface, contributing to recombination.

These electrodes are prepared by assembling polystyrene particles into molds, and subsequently infiltrating them with a TiO_2 precursor. We will discuss the methods used to fabricate titania inverse opals of a range of pore sizes, and present the advantages and disadvantages of each of them. We will also present some preliminary electrochemical analysis of these electrodes.

See more of this Session: Nanomaterials for Photovoltaics II
See more of this Group/Topical: Topical 5: Nanomaterials for Energy Applications

(173a) Silicon Nanostructures for Lithium Ion Battery Anodes

Justin Harris[1], Aaron Chockla[1], Matthew Panthani[1], Colin Hessel[1], Dariya Reid[1] and **Brian Korgel**[2], (1) Chemical Engineering, The University of Texas at Austin, Austin, TX, (2)Department of Chemical Engineering, The University of Texas at Austin, Austin, TX

Colloidal synthetic methods have been developed for a wide variety of nanocrystal and nanowire materials. Among these materials, silicon has been one of the most challenging to synthesize in in solution. The general approach is to feed molecular reactants into hot solvent environment to precipitate the nanomaterial of interest. In the case of silicon (Si), suitable reactants are needed that readily decompose in the solvent environment and the reaction must be controlled sufficiently to yield nanomaterials with the desired size and shape. Recently, we have been developing a variety of methods for the synthesis of Si nanomaterials, including amorphous and crystalline Si colloidal particles and Si nanorods and nanowires. These methods can yield significant quantities of Si nanomaterials and one potential application for them is as a replacement for carbon as the anode material in lithium ion batteries. Si has the ability to alloy with lithium ($Li_{4.4}Si$) to achieve very high charge storage capacity of as much as 4200 mAh/g, which is more than an order of magnitude higher than the storage capacity of carbon. Lithiation, however, leads to a massive volume expansion (~400%), which leads to pulverization. Si nanostructures have the ability to tolerate this volume expansion, while enabling significant lithiation. Recent data of the charge storage capacity of colloidal silicon nanoparticles and nanowires will be presented.

See more of this Session: Nanomaterials for Energy Storage III
See more of this Group/Topical: Topical 5: Nanomaterials for Energy Applications

(173b) Nano-Silicon/Graphene Composite Anode Modifications for Enhanced Performance Lithium Ion Batteries

Rhet Joseph De Guzman[1], Jinho Yang[2], Mark Cheng[2], Steven Salley[1] and K. Y. Simon Ng[1], (1) Department of Chemical Engineering and Materials Science, Wayne State University, Detroit, MI, (2) Department of Electrical and Computer Engineering, Wayne State University, Detroit, MI

Development of lithium ion batteries with improved capacity and cycle performance are in the forefront of efforts to increase the electrification of current-age vehicles. Lately, much attention has been focused in the use of silicon-based anode materials due to their low discharge potential and theoretical capacity (4,200 mAh/g) which is more than 10 fold of the traditional graphite anode. However, due to the increased accommodated lithium ions, silicon's volume varies by 400% causing pulverization and loss of electrical contact. Due to this, strategies have led to the use of nano-sized Si and incorporation of Graphene to be able to mitigate volume variation and therefore produce a stable anode material with improved capacity and cyclability. This project is aimed at the improvement of the dispersion of Si nanoparticles and the optimization of the graphene nanostructure producing a Si nanoparticle/graphene nanosheet composite anode demonstrating improved capacity retention and rate performance.

See more of this Session: Nanomaterials for Energy Storage III
See more of this Group/Topical: Topical 5: Nanomaterials for Energy Applications

(173c) Sn/SnOx /Graphene Nanocomposites: Synthesis and Anode Performance In Lithium-Ion Batteries

Mahbuba Ara[1], Dr. Steven Salley[2] and Dr.Simon Ng[2], (1)Chem. Eng., Wayne State University, Detroit, MI, (2)Chemical Engg., Wayne State University, Detroit, MI

The utilization of a Sn/SnO_x/graphene nanocomposite as anode material for lithium-ion batteries could potentially provide high reversible capacity as well as a reduction of capacity fade upon cycling, by increasing the Sn:Li_2O ratio in the anode matrix. A two step chemical synthesis route has been developed to produce a Sn/SnO_x/Graphene nanocomposite. In the first step, SnO_2 nanoparticles were prepared by a hydrolysis method and loaded on graphene nanosheets using an ultrasonic treatment to make SnO_2/graphene nanocomposites. In the second step, Sn/SnOx/graphene was prepared by the partial reduction of SnO_2/graphene in a 10% H_2 in Ar mixture at 700 °C. To get nanosized Sn/SnOx particles with size in the range of 5-10 nm, reaction time and temperature were varied to control the nucleation and to obtain a homogeneous distribution of Sn nanoparticles on the graphene sheets, while pH of the solution was optimized, as at a specific pH the surface charge of the particles affects particle organization on the surface. The small size and uniform distribution of Sn/SnOx nanoparticles are designed to reduce the volume expansion of Sn during cycling. Hence, the anode of lithium-ion battery endures lower strain and is less likely to be cracked upon cycling. Sn:C ratios of 1.5, 2.5 and 3.5 in the Sn/SnOx/graphene nanocomposites will be evaluated for a superior Li-storage performance with good cycle life and high capacity.

See more of this Session: Nanomaterials for Energy Storage III
See more of this Group/Topical: Topical 5: Nanomaterials for Energy Applications

(173d) A Facile Way to Fabricate MoO3 /Graphene Composite Anode for Lithium Ion Battery

Yun Xu, Qianglu Lin, Ling Fei, Joshua Hill, Shuguang Deng and Hongmei Luo, Chemical Engineering Department, New Mexico State University, Las cruces, NM

A Facile Way to Fabricate MoO_3 /Graphene Composite Anode for Lithium Ion Battery

Yun Xu, Qianglu Lin, Ling Fei, Joshua Hill, Shuguang Deng, Hongmei Luo

Department of Chemical Engineering, New Mexico State University, Las Cruces, New Mexico 88003

MoO_3 is capable of Li-$^+$ insertion/extraction in excess of 6 Li-$^+$ per formula unit, resulting in a significantly reversible capacity. Recently, chemically modified graphene materials have been used to form hybrid materials with SnO_2, Fe_2O_3, and Mn_3O_4 in attempts to improve electrode capacity and cycling stability. However, an anode combining MoO_3 with graphene has never been reported. MoO_3/ graphene composite was prepared through a facile two-step synthesis by homogeneous coprecipitation and subsequent reduction of the graphite oxide with hydrazine. The synergic effect of graphene and MoO_3 improves the electrochemical properties, like cycling performance and rate capability. SEM and TEM images show that MoO_3 uniformly dispersed on the surface of graphene. The synthesis approach presents a promising route to produce large amount of high performance electrode materials for Lithium ion batteries.

See more of this Session: Nanomaterials for Energy Storage III
See more of this Group/Topical: Topical 5: Nanomaterials for Energy Applications

(173e) Graphene-Encapsulated Hollow Fe3O4 Nanoparticle Aggregates As a High Performance Anode Material for Lithium Ion Batteries

Dongyun Chen[1], Ge Ji[1], Yue Ma[1], **Jim Yang Lee**[2] and Jianmei Lu[3], (1)Chemical & Biomolecular Engineering, National University of Singapore, Singapore, Singapore, (2)Department of Chemical and Biomolecular Engineering, National University of Singapore, Singapore, Singapore, (3)Chemical Engineering and Materials Science, Soochow University, Suzhou, China

Graphene-EncapsulatedHollow Fe_3O_4 NanoparticleAggregates as a High Performance Anode Material for Lithium Ion Batteries

Dongyun Chen[1,2], Ge Ji[1], Yue Ma[1], Jim Yang Lee[1] and Jianmei Lu[2]

-

[1]*Department of Chemical& Biomolecular Engineering, Faculty of Engineering, National University ofSingapore, 10 Kent Ridge Crescent, Singapore, 119260; Tel: (65) 65162186; E-mail:cheleejy@nus.edu.sg.*

[2]*Key Laboratory ofOrganic Synthesis of Jiangsu Province, Key Laboratory of Absorbent Materialsand Techniques for Environment, College of Chemical, Chemical Engineering andMaterials Science, Soochow University, Suzhou, China, 215123; E-mail: lujm@suda.edu.cn*

In this work we combined the unique propertiesof graphene sheets and a hollow assembly of nanoparticles to simultaneouslyprovide large reversible Li^+ storage capacity,good rate performance and long cycle life. Specifically graphene-encapsulatedordered aggregates of Fe_3O_4 nanoparticles withnearly-spherical geometry and hollow interior (G-HPM)were synthesized by a simple self-assembly process. TheHPM particles were first prepared by a one-pot hydrothermal synthesis. Theywere then modified by 3-aminopropyltriethoxysilane (KH550) to acquire a positivesurface charge, which enabled electrostatic assembly with negatively chargedgraphene oxide. Finally, a chemical reduction was used to reduce the grapheneoxide to graphene; finalizing the packaging of HPM particles in individualgraphene cages. This assembly could be carried out under relatively mildconditions; thereby minimizing the perturbations to the properties of HPM. Inour design, the open interior structure has adapted well to the volume changein repetitive Li^+ insertion and extraction reactions; and theencapsulating graphene connects the Fe_3O_4 nanoparticleselectronically. The composite particles showed a stable high specificreversible capacity of about 897 mA h g^{-1} which was nearly unvaryingover 50 cycles. These graphene-Fe_3O_4 composites are thereforea capable reversible Li^+ host with high storage capacity and goodcycle life which can be cycled at high rates.

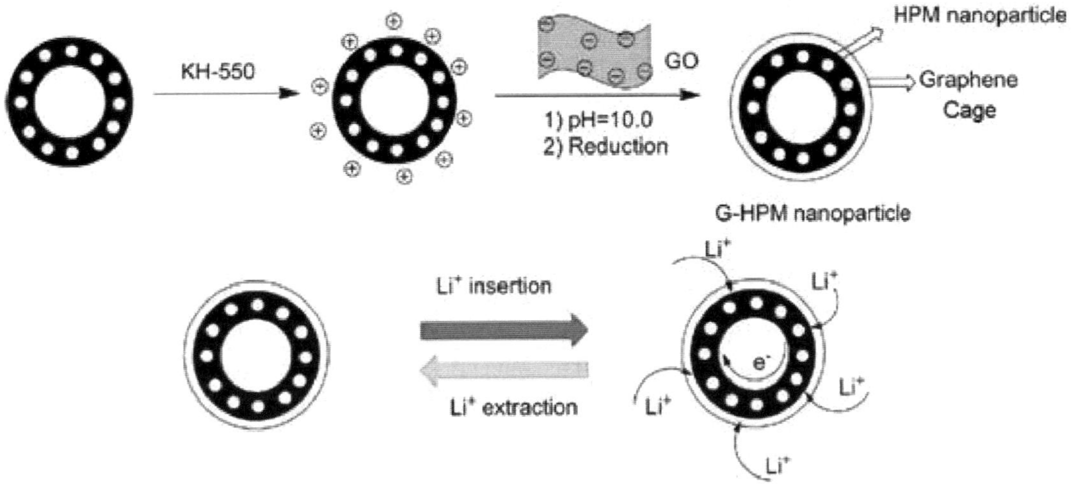

See more of this Session: Nanomaterials for Energy Storage III
See more of this Group/Topical: Topical 5: Nanomaterials for Energy Applications

(173f) An In Situ Method of Creating Metal Oxide-Carbon Nanocomposite As Anode Materials for Lithium-Ion Batteries

Zichao Yang, Jingguo Shen and Lynden A. Archer, Chemical and Biomolecular Engineering, Cornell University, Ithaca, NY

Transition metal oxides are attractive as anode materials for lithium-ion batteries (LIBs) due to their high lithium intercalation capacity and generally high natural abundances and their use on the nanoscale is being actively pursued. However, there are still challenges associated with the use of these metal oxides, including the pulverization problem and limited electrical conductivity of many metal oxides. It has been shown that these problems may be alleviated by creating metal oxide-carbon composites, which improve upon the mechanical flexibility and the electrical conductivity of the material.

There are different approaches for creating the carbon composites such as carbon nanopainting and attaching the metal oxide nanoparticles to carbon substrates. The current work describes an alternative method of creating metal oxide-carbon nanocomposites, in which the metal oxide nanoparticles are created *in situ* and embedded in a porous carbon matrix, which is made through calcination of a polyacrylonitrile-based polymer. Fe3O4 is used as a model compound.

Characterization techniques such as X-Ray diffraction, transmission electron microscopy and Raman spectroscopy were used to investigate the physical and chemical properties of the composite. Cyclic voltammetry of the composite shows lithium intercalation at 1.2V and 0.8V, which may be attributed to the reduction of Fe^{3+} to Fe^{2+} and Fe^{2+} to Fe^{0}, respectively. Very stable capacity is observed over >110 cycles at charging rate of 1C. Performance at other charging rates up to 5C is also shown and is seen to be stable. The contribution of factors including electrical conductivity, mechanical properties and porous nature of the carbon matrix in enhancing the electrochemical performance is investigated. The application of the method to other related materials such as MnO, Sn and Si is discussed.

See more of this Session: Nanomaterials for Energy Storage III
See more of this Group/Topical: Topical 5: Nanomaterials for Energy Applications

(173g) Binderless Cobalt Oxide Nanoparticle Films Through Electrophoretic Deposition, for Li-Ion Batteries

Don-Hyung Ha, Mohammad A. Islam and **Richard D. Robinson**, Materials Science and Engineering, Cornell University, Ithaca, NY

Nanoscalematerials for Li-ion battery electrodes have received great attention due totheir high surface-to-volume ratio and reduced Li-ion diffusion length whichresults in faster charge/discharge. Nanomaterialsalso show the ability to accommodate strain through volume expansion without pulverizationor capacity fading. In this work we have developed a hollow cobalt oxide NP filmfor Li-ion battery electrodes that require no additional support materials suchas carbon black or polymers to establish good electrical pathways. Thecolloidal cobalt NPs are prepared by decomposition of organometallicprecursors in non-aqueous conditions. The NP film is formed by using electrophoretic deposition (EPD). During EPD, a film isformed on both the cathode and anode but the film on the anode is slightlythicker than that of the cathode, which indicates that more particles arenegatively charged. The EPD deposited films are calcinedto form hollow cobalt oxide (Co_3O_4) NPs that act as the batteryanode.

Battery cycling of these anodes show lowdegradation and are near the theoretical capacity limits for this material. Theseresults are surprising since control experiments made from dropcastfilms form poor battery contacts under the same calcination and battery-test conditions.

See more of this Session: Nanomaterials for Energy Storage III
See more of this Group/Topical: Topical 5: Nanomaterials for Energy Applications

(174a) Polymer Solar Cells, Deconstructed

Yueh-Lin Loo, Chemical Engineering, Princeton University, Princeton, NJ

Soft-contact lamination and delamination has enabled us to construct polymer solar cells for testing, and deconstruct them subsequently for structural characterization of the active layers and electronic characterization of relevant charge transfer interfaces. We have thus been able characterize buried active layers and interfaces that are otherwise inaccessible. Structural characterization post-device fabrication and testing reveals simultaneous crystallization of the polymer donor and the electron acceptor in the once-buried bulk-heterojunction active layers to be responsible for photocurrent enhancement in these devices. Electronic characterization of the active layer-bottom electrode interface reveals an electronic band gap of 1.5 eV, attributable to the difference between the ionization potential of the polymer donor and the electron affinity of the electron acceptor. This value is significantly larger than the band gap predicted by examining the energy levels of the individual constituents, likely due to the presence of interfacial dipoles when the polymer donor and the electron acceptor are intimately mixed.

See more of this Session: Nanomaterials for Photovoltaics III
See more of this Group/Topical: Topical 5: Nanomaterials for Energy Applications

(174b) Transparent, near-Infrared Organic Photovoltaic Solar Cells for Window and Energy-Scavenging Applications

Richard R. Lunt, Dept. of Chemical Engineering & Materials Science, Michigan State University, East Lansing, MI and Vladimir Bulovic, Dept. of Electrical Engineering, Massachusetts Institute of Technology, Cambridge, MA

Building integrated photovoltaic technologies are an enticing route to capturing large areas of solar flux necessary to offset a significant portion of our non-renewable energy consumption. In this work we explore a route to high-efficiency and high-transparency solar cells that can be integrated onto window panes in homes, skyscrapers, and automobiles to enhance the functionality of already utilized transparent surfaces. We exploit spectrally-discrete excitonic features of molecular photoactive layers to position the solar cell absorption in the near-infrared (NIR), leaving the visible part of the spectrum largely unaffected. By optimizing the near-infrared optical interference generated within the transparent ITO contacts, we demonstrate planar heterojunctions with power conversion efficiencies of 1.3±0.1% and simultaneous average transmission across the visible spectrum of > 65%. This selective photoactivity in the NIR also allowed us to further optimize the architecture with the use of a transparent NIR reflector, resulting in an efficiency of 1.7±0.1%, approaching the 2.4±0.2% efficiency of the opaque control cell while maintaining high average visible transparency of > 55%. Finally we demonstrate a useful application of this technology, whereby a series of the transparent cells are monolithically integrated to power small electronic components with low-light levels. This transparent photovoltaic architecture suggests new strategies for high-efficiency power generating windows, and highlights an application benefiting from excitonic electronics.

See more of this Session: Nanomaterials for Photovoltaics III
See more of this Group/Topical: Topical 5: Nanomaterials for Energy Applications

(174c) Evidence for High-Efficiency Exciton Dissociation At Polymer/Single-Walled Carbon Nanotube Interfaces In Planar Nano-Heterojunction Photovoltaics

Geraldine LC Paulus[1], Moon-Ho Ham[1], Chang Young Lee[1], Changsik Song[1], Kourosh Kalantar-Zadeh[2], Wonjoon Choi[3], Jae-Hee Han[4] and Michael Strano[4], (1)Department of Chemical Engineering, Massachusetts Institute of Technology, Cambridge, MA, (2)School of Electrical and Computer Engineering, RMIT University, Melbourne, Australia, (3)Department of Mechanical Engineering, Massachusetts Institute of Technology, Cambridge, MA, (4)Chemical Engineering, Massachusetts Institute of Technology, Cambridge, MA

Evidence for high-efficiency exciton dissociation at polymer/single-walled carbon nanotube interfaces in planar nano-heterojunction photovoltaics

There is significant interest in combining carbon nanotubes (CNTs) with semiconducting polymers for photovoltaic applications because of potential advantages from smaller exciton transport lengths and enhanced charge separation. However, to date, bulk heterojunction (BHJ) devices with CNTs incorporated in them have demonstrated relatively poor efficiencies, and little is understood about the polymer/nanotube junction. To investigate this interface, we fabricated a planar heterojunction (PHJ) comprising well-isolated millimeter-long single-walled carbon nanotubes underneath a poly(3-hexylthiophene) (P3HT) layer. The resulting junctions display photovoltaic efficiencies per nanotube ranging from 3% to 3.82%, which exceed those of polymer/nanotube BHJs by a factor of 50-100.(1) The increase is attributed to the absence of aggregate formation in this planar device geometry. It is shown that the polymer/nanotube interface itself is responsible for exciton dissociation. Typical open-circuit voltages are near 0.5 V with fill factors of 0.25-0.3, which are largely invariant with the number of nanotubes per device and P3HT thickness. A maximum efficiency is obtained for a 60 nm-thick P3HT layer. Another PHJ described in literature using P3HT as the donor but PCBM as the acceptor (rather than SWNTs) also shows a maximum photocurrent output at a P3HT thickness of 60-65 nm, in contradiction to the expected value equal to the diffusion length of excitons in P3HT (8.5nm). We investigate the photocurrent generation in these two state-of-the-art devices by -for the first time, combining an optical T-matrix model with a Kinetic Monte Carlo Simulation. The combined model takes into account the rates of exciton generation, transport, recombination and dissociation using literature values. By including the optical, electronic and structural properties of the different materials, we are able to predict the short-circuit current of recently reported P3HT/SWNT PHJ and P3HT/PCBM PHJ solar cells from the literatureThe model demonstrates how a bulk exciton sink can explain this shifted maximum in the P3HT/SWNT case, whereas the maximum is mainly determined by PCBM interdiffusing in P3HT in the P3HT/PCBM case. Based upon the results of this model it will be possible to more intelligently design polymer hybrid solar cells (both planar and bulk) and optimize them towards higher efficiencies.

Both the experimental as well as the simulation platform are promising for further understanding the potential role of polymer/nanotube interfaces for photovoltaic applications.

1. Ham MH, Paulus GLC, Lee CY, Song C, Kalantar-zadeh K, Choi W, Han JH and **Strano MS**: Evidence for High-Efficiency Exciton Dissociation at Polymer/Single-Walled Carbon Nanotube Interfaces in Planar Nano-heterojunction Photovoltaics. ACS NANO, 4 (2010) 6251-6259

See more of this Session: Nanomaterials for Photovoltaics III
See more of this Group/Topical: Topical 5: Nanomaterials for Energy Applications

(174d) Combinatorial Organic Photovoltaic Microfluidic Device

Michael C. Fechtmann and Shahab Shojaei Zadeh, Mechanical and Aerospace Engineering, Rutgers, The State University of New Jersey, Piscataway, NJ

The introduction of conjugated organic semiconductors to the field of photovoltaics has dramatically increased the number of materials available from which to construct solar cells. The theory of operation of organic solar cells based on the highest occupied molecular orbital (HOMO) of the electron donor and the lowest unoccupied molecular orbital (LUMO) of the electron acceptor provides a guideline for the synthesis of ideal n-type and p-type organic semiconductors which have the potential to achieve a conversion efficiency of over 10%. From an experimental standpoint, however, many material combinations go untested. Also, cell efficiency declarations are generally not very repeatable and are based on the best of a group of cells. This is mainly due to the fact that factors such as sealing and thickness control in these devices are not easily achievable. In this work, a microfluidic device has been designed to facilitate solar cell construction by making multiple n-type and p-type organic semiconductor combinations simultaneously. This design consists of multiple split chambers in which an n-type organic semiconductor is deposited in one half and then a p-type organic semiconductor is injected to fill the other half. This creates an interface between the two materials. Each cell has its own ITO coated polymer electrode and a metallic counter electrode which allow for measurements. A solar simulator is used to determine and compare the energy conversion efficiency of each cell. Dye sensitized solar cells can also be constructed and tested in a similar fashion, with electrolyte flowing through the microchannels to wet an electrode and dye combination in the chambers.

See more of this Session: Nanomaterials for Photovoltaics III
See more of this Group/Topical: Topical 5: Nanomaterials for Energy Applications

(174e) Structure and Properties of Porous Poly(3-hexylthiophene) Gel Particles In Aqueous Dispersion

Jeffrey J. Richards and Danilo C. Pozzo, Chemical Engineering, University of Washington, Seattle, WA

The demand for thin-film solar cells has increased significantly because modules can be inexpensively produced using scalable roll-to-roll processing techniques. We have developed aqueous dispersions of colloidal poly(3-hexylthiophene) (P3HT) gel particles that can serve as solution-processable inks for active layer materials in polymer solar cells. Bulk P3HT organogels have been emulsified into water to produce sub-micron gel particles that are stabilized with surfactant. The emulsification process produces stable dispersions of porous P3HT particles that maintain a high crystallinity, high porosity and the networked structure of the original organogel. The optical characteristics, structure and stability of the resulting gel particles have been thoroughly assessed using spectroscopy, neutron scattering, DLS and electron microscopy. It is shown that the P3HT particles retain the crystallinity and high degree of intrachain order with no evidence of doping. Simple processing is demonstrated via spray-coating of the dispersions. Thin-films have been prepared and their structure and properties have been characterized. The use of gel particle dispersions opens up a new avenue to produce thin-films for polymer solar cells and organic electronic applications while minimizing the need for post-processing.

See more of this Session: Nanomaterials for Photovoltaics III
See more of this Group/Topical: Topical 5: Nanomaterials for Energy Applications

(174f) Core-Shell Nanostructures of Organic-Inorganic Nanomaterials for Hybrid Photovoltaics

Chinedum Osuji[1], Shanju Zhang[2], Candice Pelligra[2] and Lisa Pfefferle[2], (1)Chemical Engineering, Yale University, New Haven, CT, (2)Chemical and Environmental Engineering, Yale University, New Haven, CT

Ordered bulk heterojunctions in hybrid photovoltaics show great promise towards high efficiency devices. We report ordered core-shell nanostructures of organic-inorganic nanohybrids processed from solution. Inorganic semiconductor nanowires are prepared by facile solvothermal synthesis. High aspect ratio of the nanowires is controlled by optimizing reaction parameters. Core-shell nanohybrids are synthesized by side-on grafting conjugated polymers onto the inorganic nanowires. We demonstrate that high aspect ratio nanowires and their core-shell nanohybrids spontaneously form liquid crystalline phases in solution. These systems show isotropic, bi-phasic and nematic phases with an increase of concentration, which is in reasonable agreement with Onsager's theory for rigid rodlike liquid crystals. Suspensions are readily processed to produce ordered aligned films with large-area monodomains. It is found that the conjugated polymer forms an ordered crystalline layer on the nanowire, as opposed to its noncrystalline bulk counterpart. The side-on grafting of polymers on the nanowires results in aligned chain-extended backbones along the nanowire long axis in contrary to the chain-folded backbones by end-on grafting of polymers. Decreasing the nanowire diameter tends to increase the chain-extended conjugated length. Such uniaxially aligned, fully chain-extended, highly ordered crystalline polymer nanostructures across interface could significantly enhance electron transfer, charge transport and collection. The corresponding optoelectronic properties are addressed.

See more of this Session: Nanomaterials for Photovoltaics III
See more of this Group/Topical: Topical 5: Nanomaterials for Energy Applications

(174g) Synthetic Iron Pyrite (FeS2) Nanocrystals for Use In Hybrid Organic-Inorganic Solar Cells

John Bae[1], Beau J. Richardson[2], Leize Zhu[2] and Qiuming Yu[1], (1)Chemical Engineering, University of Washington, Seattle, WA, (2)Dept. of Chemical Engineering, University of Washington, Seattle, WA

One of thegreatest sources of energy comes from solar radiation but the high costassociated with fabricating and maintaining silicon solar cells prevents itsspread in popularity. Low-cost and highly efficient photovoltaic devices haveto be developed in order to significantly increase the proportion of solarenergy to the total energy used.† A material that is promising in this aspectis iron pyrite (FeS_2, commonly known as fool's gold), which is abundantas well as environmentally benign with a useful band gap of ~0.95 eV and a highoptical absorption coefficient ($\alpha > 10^5$ cm^{-1}, 2 orders of magnitude greater thancrystalline silicon).† Because of this high optical absorption coefficient,iron pyrite can absorb most incident light with much less material than silicon-basedphotovoltaic devices. †Here we report a simple method of synthesizingcrystalline FeS_2 (pyrite) nanocrystals via a hydrothermal procedure.†In particular, we aim to control the size and morphology of nanocrystals inorder to investigate the surface and interfacial properties on the performanceof hybrid organic-inorganic photovoltaic devices.

Iron pyritenanocrystals were synthesized using a hydrothermal method. †Typically, poly (vinylpyrrolidone)(PVP) solution was mixed with poly(vinyl alcohol) (PVA) solution and $FeCl_2\Sigma 4H_2O$ followed by the addition of NaOH and sulfurpowder.† After stirring the mixture for 30 minutes, the Teflon liner containingthe reaction mixture was placed in a stainless steel autoclave and reacted at170-220∞C for 12-24 hours.† The reactor was coolednaturally to room temperature. †Once cooled, the products were washed withdeionized water and absolute ethanol several times, and dried under vacuum. †Thepyrite phase was confirmed by powder XRD as well as high resolution TEM andselected area electron diffraction (SAED). †The size and morphology weredetermined by SEM. †Results indicate that the molecular weight of surfactants(PVP and PVA), their concentration and ratio, as well as the concentration ofNaOH play important roles in controlling the morphology, i.e., cubic oroctahedral shape, of the resulting nanocrystals. †The reaction mechanisms,especially the interaction of surfactants with different crystalline facets guidingthe final morphology will be discussed. †Furthermore, the synthesizednanocrystals were blended with P3HT polymer and hybrid solar cells were made. †Therelationship between the size and morphology of nanocrystals and the solar cellpower conversion efficiency were obtained. †The effect of crystalline facets onthe charge generation and separation at the polymer-nanocrystal interface willbe discussed.

See more of this Session: Nanomaterials for Photovoltaics III
See more of this Group/Topical: Topical 5: Nanomaterials for Energy Applications

(244a) Carbon Negative Biomass Refining System Based On Alkaline Hydrothermal Treatment Technology

Thomas Ferguson and **Ah-Hyung Alissa Park**, Earth and Environmental Engineering and Chemical Engineering, Columbia University, New York, NY

A novel synthesis scheme of hydrogen from biomass is developed based on alkaline hydrothermal treatment. The proposed alkaline hydrothermal treatment is a less-studied method of biomass conversion but one with great transformational potential since it can be achieved with much higher overall conversion, the synthesized gaseous product is carbon-free and an option of carbon storage is inherently incorporated into the reaction scheme. The central concept pursed in this study is the use of a strong base in removing carbon from the C, H, O equilibrium that can be produced in mixtures of biomass and steam. The previous research has demonstrated several examples of interactions between simpler biomass (i.e., glucose and cellulose) and a strong base (i.e., NaOH) that resulted in hydrogen production with high purity (e.g., less than 10 ppm of CO) at relatively low reaction temperature and pressure (i.e., 473-773K and 1 atm). Therefore, the product gas can be directly applied to a PEM fuel cell system without further conditioning. Alkaline biomass reforming is driven in part by the heat of carbonation and the main concerns regarding this technology would be the method of hydroxide recovery. The reaction kinetics and conversion of the proposed biomass conversion system are investigated for the varied reaction temperature, biomass-hydroxide ratio, catalysts, partial pressure of H2O, etc. A series of analytical techniques such as GC, MS, NMR, ATR-FTIR and XRD are employed to fully identify each individual species in the product stream and to provide insights into this new reaction pathway toward biomass-driven hydrogen.

See more of this Session: Nanomaterials for Hydrogen Production and Fuel Cells I
See more of this Group/Topical: Topical 5: Nanomaterials for Energy Applications

(244b) Inorganic Catalytic Nanofibers for the Production of Hydrogen Via Alkaline Hydrolysis of Biomass

Nathaniel S. Hansen[1], Daewon Cho[1], Thomas Ferguson[2], Ah-Hyung Alissa Park[3] and Yong L. Joo[1], (1)School of Chemical and Biomolecular Engineering, Cornell University, Ithaca, NY, (2)Earth and Environmental Engineering, Columbia University, New York, NY, (3)Earth and Environmental Engineering and Chemical Engineering, Columbia University, New York, NY

Here we propose to develop a farm-scale biorefining system that can convert cellulosic biomass into useable hydrogen gas. In particular, we focus on the development of cost-effective procedures for biomass pretreatment and catalysts for alkaline hydrothermal treatment, which is less studied as compared to conventional gasification and pyrolysis but with great potential. While we have made progress in chemical routes to directly incorporate catalysts into cellulose, one-dimensional materials also have been investigated to support catalysts thereby maintaining the nanoparticles' high surface area to volume ratio while inhibiting their aggregation tendency. Electrospinning in particular has garnered significant interest due to the cost effective nature and diversity of materials/morphologies available. In this study monoaxial and coaxial inorganic silica nanofibers with in-situ formed iron or nickel catalytic nanoparticles have been fabricated through sol-gel chemistry with a metallic precursor and subsequent thermal treatment with coaxial nanofibers deliberately tuning the catalyst location to the surface of the nanofiber. Further, a purely water based approach has been used by homogeneously binding high concentrations metallic and ceramic precursors to a polymer carrier. Electrospinning and subsequent thermal treatment generates nanoscale fiber diameters of morphologies consisting of pure metal nanofibers or discrete metal domains in a ceramic substrate through monoaxial electrospinning, or purely metallic shells surrounding a purely ceramic core or discrete metal domains tuned toward the surface of the nanofiber through coaxial electrospinning. The morphology, oxidation state and crystal location are studied using scanning electron microscopy (SEM), transmission electron microscopy (TEM) and X-ray diffraction (XRD). The resulting nanofibers were applied as a catalytic system for the alkaline hydrolysis of biomass to selectively produce hydrogen with minimal carbon monoxide or dioxide byproducts. Our results indicate that by tuning the catalyst type, catatlyst morphology, substrate composition, reaction temperature and heating rate, nearly 100% conversion can be achieved at low temperatures and low pressures with a mole fraction of hydrogen greater than 97%. With these preliminary results, this system is especially exciting because it lends itself easily to the potential on site, cost effective generation of useable fuel gas from a biomass feed stock.

See more of this Session: Nanomaterials for Hydrogen Production and Fuel Cells I
See more of this Group/Topical: Topical 5: Nanomaterials for Energy Applications

(244c) Photocatalytic Hydrogen Production Using Coated TiO_2 Substrate

Njideka H. Okoye[1], Pedro E. Arce[1] and Dennis B. George[2], (1)Chemical Engineering, Tennessee Technological University, Cookeville, TN, (2)Center for the Management, Utilization, and Protection of Water Resources, Tennessee Technological University, Cookeville, TN

This study provides insight regarding photocatalytic hydrogen production by a transparent thin film of TiO_2 photocatalyst coated on a glass substrate doped with Platinum. The experiment was designed into two parts: The first part was on the manufacturing and physical characterization of titanium dioxide surface coating and the second part focused on hydrogen production. The TiO_2 thin coatings were characterized by using different analytical methods, like X-ray diffraction (XRD), scanning electron microscope (SEM), transmission electron microscope (TEM) and UV-Vis spectrophotometer. The effects of some hydrogen production variables such as volume percentage composition of ethanol and surface area of coated substrate were investigated over a period of two hr interval. Effectiveness of coatings on the substrate was tested by reusing the substrate with new solution of ethanol. From the analysis performed it was observed that hydrogen production rate significantly increased as percentage ethanol composition increased. Similar trend was also observed with respect to the increase surface area. The catalyst surface was also obsevered to be deactivated with further reusing.

See more of this Session: Nanomaterials for Hydrogen Production and Fuel Cells I
See more of this Group/Topical: Topical 5: Nanomaterials for Energy Applications

(244d) Ordered Mesoporous TiO2/ZnO Material for Hydrogen Production

Menan Hussein and Rajesh Shende, Chemical and Biological Engineering, South Dakota School of Mines and Technology, Rapid City, SD

Ordered mesoporous Zr-doped TiO_2/ZnO material is currently being investigated for band gap tuning and higher loadings of photo-sensitizer. Utilization of such mesoporous materials thereby offers an opportunity to sensitize semiconductor with higher sensitizer loadings leading to effective charge separation and higher photo-energy conversion. It may be possible to achieve the band gap tuning as well as higher sensitizer loadings through the formation of Zr-doped mesoporous TiO_2/ZnO networks by the sol gel method. Ordered mesoporous Zr-doped TiO_2/ZnO materials as thin-film and powder were synthesized using hexadecylamine (HDA) as the surfactant, titanium ethoxide and $ZnCl_2$ precursors. The gels were aged and calcined for 5 hours in nitrogen environment at 500 ^{o}C. H_2 generation experiments were performed using Zr-doped TiO_2/ZnO powder in quartz reactor cell (7 ml) containing water and NaI mixture. The cell was illuminated with 500 W Xe lamp with different power density. The gas samples were withdrawn periodically and analyzed using GC provided with thermal conductivity detector. SEM/TEM analysis of mesoporous TiO_2/ZnO material and H_2 generation from photocatalytic water-splitting reaction will be presented.

See more of this Session: Nanomaterials for Hydrogen Production and Fuel Cells I
See more of this Group/Topical: Topical 5: Nanomaterials for Energy Applications

Nanostructured ZnO Thin Film for Photoelectrochemical Splitting of Water: A Study on Role of Electrolytes

Pushpendra Kumar[1], *Vidhika Sharma*[1], *Nirupama Singh*[1], *Anjana Solanki*[1], *Sumant Upadhyay*[1], *Surbhi Chaudhary*[1], *Vibha R Satsangi*[2], *Sahab Dass*[1] *and Rohit Shrivastav*[1]
([1])Department of Chemistry, Dayalbagh Educational Institute, Agra, India,
([2]) Department of Physcics and Computer Science, Dayalbagh Educational Institute, Agra, India

Abstract: The photoelectrochemical production of hydrogen from water using solar energy has drawn considerable attention due to the importance of using hydrogen as a clean and renewable source. The effectiveness of such a photoelectrochemical cell in converting radiation energy to electricity depends upon the efficiency with which light is absorbed by the photoanode, the number of e^- - h^+ pairs generated at the photoanode and their separation. In this study, employing nanostructured ZnO thin films supported on conducting substrate as working electrode in PEC splitting of water, the role of different electrolyte combinations have been investigated. Besides, the ZnO thin films prepared by sol-gel spin coating were also subjected to extensive characterization of the structural, morphological, electrical, optical and electrochemical properties to assess their effect on photoelectrochemical behaviour of ZnO thin films. X-Ray diffraction pattern confirmed the prominent evolution of thermodynamic stable hexagonal wurtzite ZnO having band gap 3.16eV. The films were subjected to SEM and AFM to investigate surface morphology and surface topography, respectively. As synthesized films were converted into working electrode and employed in photoelectrochemical cell under varying electrolytic conditions. The films offered high photovoltage and short circuit current. Photocurrent of 1mA was recorded under short circuit condition in aqueous electrolyte at 13pH. Some of the electrolyte combination variants investigated are pH, addition of sacrificial donor, and hole scavenger. It was observed that such additives to the electrolyte play an important role in reduction of electron - hole recombination which in turn enhanced the overall photocurrent.

Introducton

Photoelectrochemical (PEC) splitting of water into H_2 and O_2 using semiconductors has received much attention, especially for its potential application in production of a clean energy carrier from solar light energy (1,2). The PEC effect of the semiconductor originates from creation of the electrons and holes after absorption of light. Since the report of Fujishima and Honda (3) on water-splitting by PEC cells, numerous efforts have been made for the construction of efficient and practical semiconductor system for this purpose. In recent years the possibility of using nanostructured ZnO for PEC splitting of water to produce hydrogen has been explored by some researchers (4-6). Despite its large band gap it is a preferred material because of its extraordinary stability in aqueous environment. The energy levels and band gap energy of ZnO are quite similar to TiO_2, which is the most extensively investigated material for PEC splitting of water, so far. Further, electron mobility in ZnO is much higher compared TiO_2 and it can be an advantage for fabricating current generating devices such as PEC/solar cells (6).

In present investigation, nanostructured ZnO was synthesized by sol-gel spin coating method and employed as a photoanode in PEC cell and authors also investigated effects of electrolyte environment on the process.

Experimental

Zinc acetate dehydrate (ZAD), $Zn(COOCH_3)_2.2H_2O$, was used as the starting compound to prepare ZnO thin films. To a solution of ZAD prepared in 2-methoxyethanol, monoethanolamine (MEA) (used as stabilizer) was added at room temperature. The molar ratio of MEA to ZAD was kept to 2. The mixture was stirred for 1 h at 50-60 °C till a clear and transparent solution was obtained which was spin coated on TCO glass plates (having SnO_2: F overlayer on one side) at 3000 rpm for 20 seconds.). Coated films were dried on hot plate at 60°C for 10 min., preheated at 250°C for 30 min., and then finally annealed at 600°C for 1h. The films, thus obtained were subjected to various physico-chemical characterizations. Crystal phase and particle size analysis were made by X-ray diffractometer. Band gap was determined by UV-Visible spectrophotometer. Surface analysis was done by observing AFM and SEM images. Synthesized ZnO thin films were also converted into electrode by connecting it to copper wire by silver paste and employed as working electrode in PEC cell.

Results and discussion

In this study, ZAD, was the starting material for the growth of ZnO thin films. A stable sol was prepared by dissolving ZAD in 2- methoxyethanol followed by the addition of monoethanolamine. Here MEA acts as complexing agent and also increases the pH, which promotes the formation of ZnO. In the ensuing sequence of reactions, three nucleophilic species (MEA, HO^- and CH_3COO^-) compete for the Zn^{2+} Lewis acid center. The attack of an HO^- group leads to the formation of small zinc-oxo-acetate oligomers, which are probably evolved in the initial stage from gradual complexes during aging. The

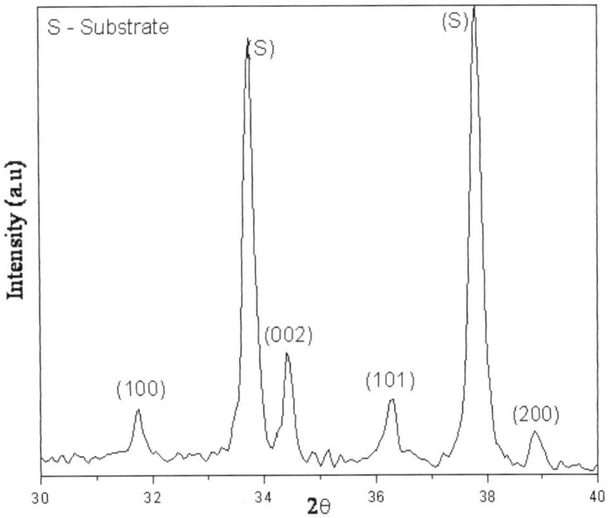

Figure 1: X-ray diffraction pattern of ZnO thin film

progressive condensation of the hydrolyzed moieties then gives rise to colloids or precipitates.

Fig. 1 is the observed diffraction pattern of ZnO thin film that suggests dominant evolution of wurtzite ZnO. Presences of multiple peaks suggest polycrystalline nature of film. The high intensity of

(002) peak at 34.42^0 shows the preferential growth along c-axis. The other peaks at 2θ angle 31.7, 36.3 and 38.7 correspond to diffraction from planes (100), (101) and (200), of hexagonal wurtzite ZnO. Thermodynamically ZnO is more stable with the wurtzite phase due to its ionicity that resides exactly at the borderline between the covalent and the ionic materials (5). The c-axis oriented preferential growth of crystallites can be understood from the fact that the value of surface energy of ZnO (002) plane is reported to be minimum at the growth stage (7). The XRD data was utilized to evaluate average crystallite size by Debye Scherrer calculation (8). The evaluated crystallite size was 52 nm.

Fig. 2 (A) shows the absorption curve of the films. The absorption at higher wavelengths in the visible region is low but at lower wavelength (< 400nm) an intense absorption threshold can be seen. Fig. 2 (B) depicts the plot of $(\alpha h\upsilon)^2$ vs. hυ from which the calculated optical band gap was 3.13eV.

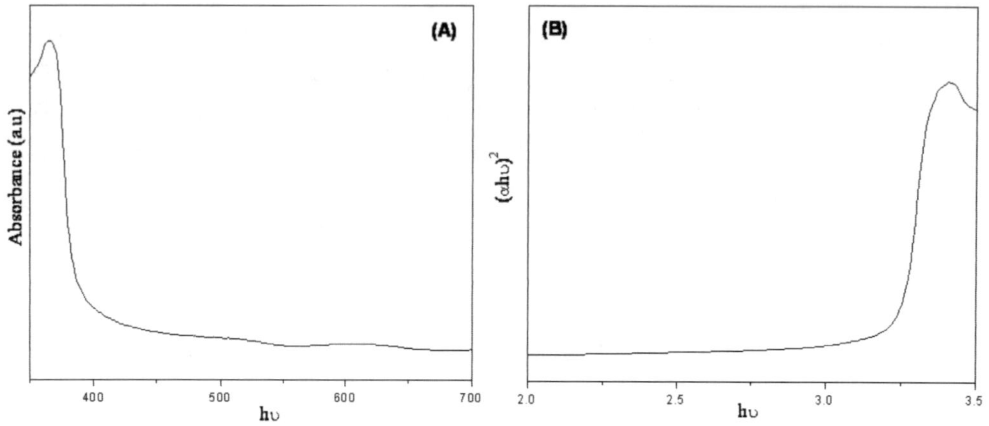

Figure 2: Plot of: (A) - Absorption vs. wavelength , (B) - $(\alpha h\upsilon)^2$ vs. hυ

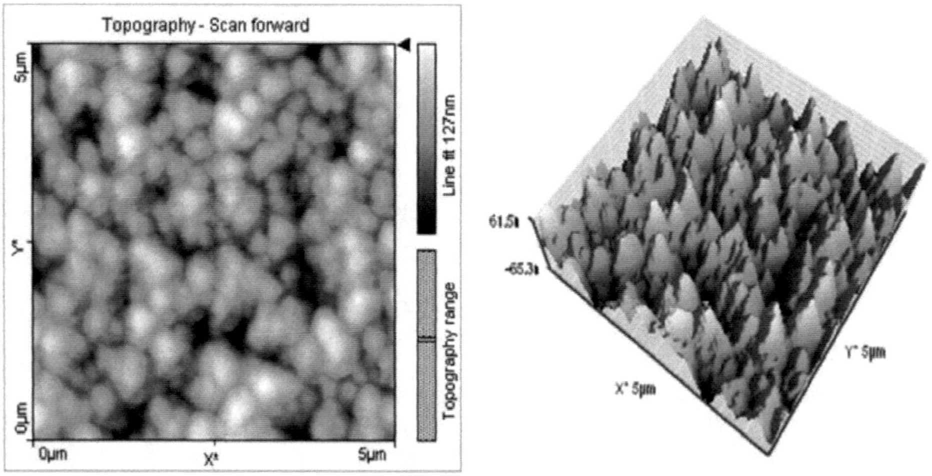

Figure 3: AFM image of ZnO thin film

Observed AFM image of ZnO thin film is shown in Fig 3. It is clear that the films are homogeneous, free from any major structural defects, compact, uniform and void free. The SEM image (Fig 4) also shows formation of fine granules of crystals and supports the morphology as depicted by AFM.

Figure 4: (A) SEM Image, (B) Cross-sectional SEM Image of ZnO thin film

To investigate PEC behaviour of ZnO thin films, films were converted into electrode and employed as working electrode in PEC cell. The current – voltage (I-V) curves were recorded under darkness and illumination using 150 W Xenon arc lamp. Current voltage curves were recorded in different electrolytes, in order to analyse the effect of electrolytic conditions on photoelectrochemical behaviour of ZnO thin film. Diffrent electrolytic combinations and conditions used in the study are given in table 1. The observed current – potential I – V curves of ZnO films demonstrated a typical feature of n-type semiconductor electrodes.

From the Fig 5 it can be seen that pH of electrolyte plays an important role on the PEC behaviour of semiconductor. At pH 13, short circuit current of 1.00 mA was recorded which decreased to 0.16 mA on decreasing the pH to 11. This shows that electrochemical potential of electrolyte has marked effect on I - V behaviour of ZnO thin film. A high short circuit current at higher pH suggest efficient separation of electrons – holes under illumination but at lower pH, electron – hole recombinations probably starts to dominate which results in poor short circuit current. The open circuit potential, which is one of the most important parameter of PEC cell, is the potential difference between the working electrode and reference electrode at equilibrium and represents the contribution of light towards the minimum voltage needed for water splitting potential.

Table 1: Electrolyte type and conditions of electrolyte used for PEC study

Electrolyte Type	Electrolyte Combination	Open Circuit Potential
A	0.1M NaOH	-722
B	0.001M NaOH	-580
C	0.1M NaOH+ 0.1M Na2SO4	-697
D	0.1M NaOH + 20 mM KI	-728
E	90% 0.1M NaOH + 10 % Methanol	-796
f	90% 0.1M NaOH + 10 % Glycerol	-889

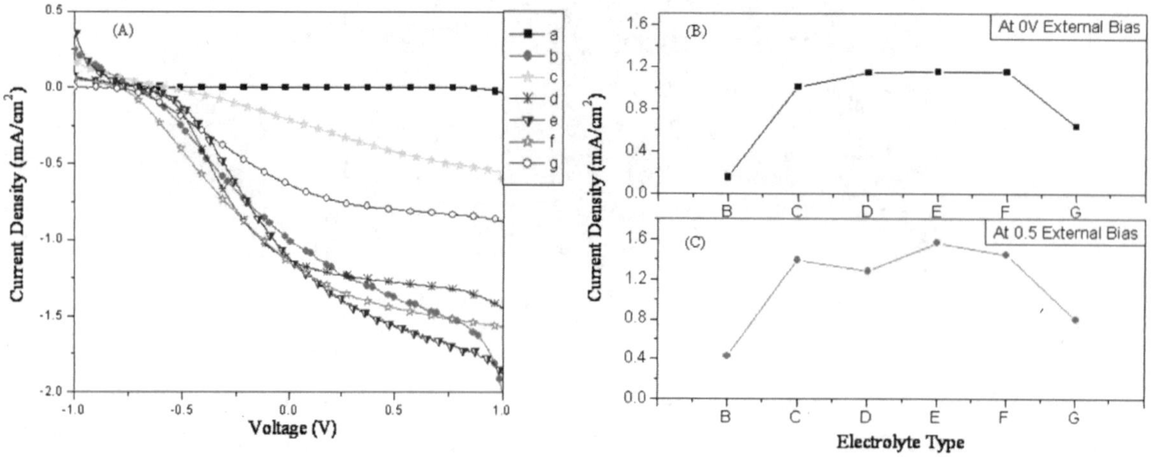

Figure 5: [A] I-V Curves recorded for ZnO thin film in different electrolytic combinations, a: Curve under darkness; b,c,d,e,f,g: under illumination in different electrolye combination (see table 1)[B] Photocurrent Density at zero bias (J_{sc}) and (C) Photocurrent Density at 0.5 V bias.

Higher band gap materials are useful for water splitting as these can supply Voc exceeding 1.23 V (water splitting potential). Values of open circuit potential (V_{oc}) observed in this study are given in Table 1 from which it can be seen that, on decreasing pH of the solution, V_{oc} decreased. The value of V_{oc} also represents the potential difference between redox potential of semiconductor and redox potential of electrolyte. In this particular case, the redox potential of electrode is same while redox potential of electrolyte is changed by varying the pH of electrolyte. To increase the ionic strength of electrolyte, 0.1 M Na_2SO_4 was added to electrolyte, keeping the pH constant and its effect on I-V curve can be seen in Fig 5. Short circuit current of 1.15 mA/cm^2 was recorded which indicate better charge separation and transportation of charge carriers. On increasing ionic strength of electrolyte, conductivity increases which helps in rapid transportation of charge carriers. However, the effect reaches to saturation earlier than plane NaOH of pH 13 and steady current was observed at lower potential. From the Fig 5 (C) it can be observed that at 0.5V external bias, the current density is lower than NaOH. Effect of KI solution (in electrolyte) on PEC response was also investigated and is shown in Figure 5 (A). A short circuit current of 1.15 mA/cm^2 was recorded. The interesting observation with KI was that there is almost no current saturation and current density of 1.57 mA/cm^2 was recorded at 0.5 V external bias. It is likely that I^- act as a strong hole scavenger which helps in electron hole separation resulting in better current density.

In photoelectrochemical reaction, electron hole recombination loss is major practical problem. When light strikes on semiconductor surface, electron – hole pairs generate and in the presence of depletion layer, they get separated. Electrons are driven to bulk and holes migrate to the solid – electrolyte interface where holes are consumed by a solution reductant. Therefore, electrolyte solution plays an important role in electron-hole separation. Addition of alcohol is reported to help in reduction of electron-hole recombination losses. In this study methanol and glycerol were added to NaOH and their effect on I-V curves was recorded. It was observed (Fig. 5(A)) that on addition of ethanol, current density increased and short circuit current of 1.15 mA/cm^2 was recorded. The marked effect of this was seen on V_{oc} which shifted to 0.79 V. This suggests that addition of alcohol not only helps to minimize electron-hole recombination losses but also modify the energy levels of electrolyte compared to O_2/H_2O energy level. However, addition of glycerol reduces the current density. If we see Fig 5 (A), very low current density was recorded but it produced very high open circuit potential, 0.88 V, which indicates that addition of alcohol produce very high photovoltage but due to high viscosity, the transporation of charge carriers was very low which reflected in terms of low current density.

Acknowledgments

The work has been financially supported by DAE – BRNS, Govt. of India and UGC, Govt. of India in the form of R & D projects sanctioned to RS.

References

1. Nowotny J., Sorrell C. C., Sheppard L. R and Bak T. (2005), "Solar-hydrogen: Environmentally safe fuel for the future", *International Journal of Hydrogen Energy,* 30, 521 – 544.
2. Krishnan R.(2007), "Hydrogen generation at irradiated oxide semiconductor–solution Interfaces , *J Appl Electrochem,* 37, 765–787.
3. Fujishima A. and Honda K. (1972), "Electrochemical photolysis of water at a semiconductor electrode" *Nature,* 238, 37-38.
4. Sharma V., Kumar P. Shrivastava J. Solanki A., Satsangi V. R., Dass S. and Shrivastav R. (2011), "Vertically aligned nanocrystalline Cu–ZnO thin films for photoelectrochemical splitting of water", *J Mater Sci.*, 46 (11), 3792-3801.
5. Xizhe L., Yanhong L., Hong L., Yuzun F., Zhexun Y., Yuan L., Liquan C., and Qingbo M. (2007), "Room temperature fabrication of porous ZnO photoelectrodes for flexible dye-sensitized solar cells", *Chem. Commun.*, 2847–2849.
6. Xunyu Y., Abraham W., Gongming W., Alissa S., Robert C. F., Fang Q., Jin Z. Z., and Yat L. (2009), "Nitrogen-Doped ZnO Nanowire Arrays for Photoelectrochemical Water Splitting", *Nano Letters,* 9 (6), 2331-2336.
7. Lee H. S., Lee J. Y., Kim T. W., Kim D. W and Cho W. J. (2004), " Formation mechanism of preferential c-axis oriented ZnO thin films grown on p-Si substrates", *J. Mater. Sci.*, 39, 3525.
8. Gupta M., Sharma V., Shrivastava J., Solanki A., Singh A. P., Satsangi V. R., Dass S. and Shrivastav R. (2009), " Preparation and characterization of nanostructured ZnO thin films for photoelectrochemical splitting of water", *Bull. Mater. Sci.*, 32, 23-30.

(319a) Enhanced Dehydrogenation Kinetics Via Thermal and Carbon Dioxide Treatments

Jae W. Lee[1], Junshe Zhang[1], Yu Zhao[2] and Daniel L. Akins[2], (1)Department of Chemical Engineering, The City College of New York, New York, NY, (2)Department of Chemistry, The City College of New York, New York, NY

Ammonium borane (AB, NH_3BH_3) is an attractive on-board, solid-state hydrogen storage material, because it has a hydrogen density of 19.6 wt. %. Hydrogen release from AB can be achieved through three approaches, including thermolysis, hydrolysis, and methanolysis. In specific, the thermolysis of solid AB includes three consecutive steps, occurring around 110, 150, and > 500 °C, with about 6.5 wt. % of hydrogen, regarding the mass of AB, desorbing in each step. Slow hydrogen release at temperatures around 85 °C makes pristine AB infeasible as an on-board hydrogen storage medium because most of polymer electrolyte membrane fuel cells operate at this temperature. Several approaches have been discussed to enhance the dehydrogenation kinetics of AB such as compositing AB with mesoporous materials, dispersing AB in ionic liquids, and adding transition metal catalysts or chemical promoters. One major drawback associated with the first two approaches is the low hydrogen storage capacity, as a result of the mass fraction of mesoporous materials and ionic liquids being usually very high. Using transition metals or promoters adds up the cost of H_2 storage system. Therfore, the research efforts have continuously made to find cheap and effective promoters for hydrogen release without negatively impacting the storage capacity. This work presents two methods for promoting hydrogen release using thermal treatments of AB at 80 °C and CO_2 pretreatments of AB at 60 - 80 °C. These two pretreatments do not require any expensive materials to enhance the kinetics and not reduce the H_2 storage density. Within an hour, hydrogen release goes up to 7.1 wt% and 8.33 wt% of AB in the thermal and CO2 pretreatment. We will propose possible mechanisms of the fast H_2 release using Raman, IR, and NMR spectroscopic methods.

See more of this Session: Nanomaterials for Hydrogen Production and Fuel Cells II
See more of this Group/Topical: Topical 5: Nanomaterials for Energy Applications

(319b) Nano-Engineered Proton and Electron Conducting Materials for Fuel Cell Applications

Chau Tran and **Vibha Kalra**, Chemical and Biological Engineering, Drexel University, Philadelphia, PA

We study the development of nano-engineered electrode/catalyst assembly for proton exchange membrane (PEM) fuel cells using nanofibers (diameter ~50-500 nm). In particular, this work focuses on controlling the nanoscale assembly of proton and electron conducting media within porous nanofiber mats. The aim is to develop catalyst supports (for PEM fuel cells) with well-defined triple phase boundary, necessary for efficient transport of electrons, protons and oxygen gas to the catalyst sites for the oxygen reduction reaction to take place at the anode.

Nanofibers are fabricated via a process called electrospinning that uses strong electric field to elongate a polymer solution or melt jet to form ultrathin fibers. Solutions of Nafion, the proton conducting ionomer in state of the art fuel cells, were not electrospinnable by themselves due to the formation of aggregates in solution owing to electrostatic interactions. Relaxation times and extensional viscosities of different Nafion solutions were measured to quantify the solution elasticity and to understand its correlation with Nafion electrospinnability. A methodology to enhance the extensional viscosity was developed that allowed us to successfully electrospin Nafion nanofibers. Scanning electron microscopy (SEM) was conducted to characterize the nanofiber diameter and morphology for different process parameters and solution compositions. Impedance spectroscopy has been conducted and Nafion nanofibers exhibit lower impedance or higher proton conductivity than bulk Nafion films. This work was followed by incorporation of electron conducting media within the Nafion nanofiber mats and transmission microscopy results exhibiting the nanoscale assembly of the two media will be presented.

See more of this Session: Nanomaterials for Hydrogen Production and Fuel Cells II
See more of this Group/Topical: Topical 5: Nanomaterials for Energy Applications

(319c) High Active PdFe-Nanoleave Catalyst for Oxygen Reduction Reaction In Alkaline Media

Zhiyong Zhang[1], Wenzhen Li[1], Karren L. More[2], Kai Sun[3] and Zili Wu[4], (1)Chemical Engineering, AICHE member, Houghton, MI, (2)Metals and Ceramics Division, Oak Ridge National Laboratory, Oak Ridge, TN, (3)Department of Materials & Engineering, University of Michigan, Ann Arbor, MI, (4) Chemical Science Division and Center for Nanophase Materials Sciences, Oak Ridge National Laboratory, Oak Ridge, TN

Abstract: Novel PdFe-nanoleaves (NLs) have been preparedthrough a wet chemistry-based solution phase reduction synthesis route.High-resolution transmission electron microscopy (HR-TEM) and scanningtransmission electron microscopy (S/TEM) coupled with high-spatial-resolutioncompositional analysis clearly show this newly-developed structure is assembledfrom Pd-rich nanowires (NWs) surrounded by Fe-rich sheets. The Pd-NWs have adiameter in range of 1.8-2.3 nm and large electrochemical surface area of>50 m^2/g. Their length (from 30 to 100 nm) and morphology can betuned through altering the nanostructure synthesis conditions and Fe amount inthe nanoleaves. The side surfaces of Pd-NWs observed by HR-TEM arepredominantly Pd (111) facet, while the tips and ends are Pd (110) and Pd (100)facets. By etching away the enveloping Fe-rich sheets using an organic acid,the Pd-rich NWs are exposed on the surface of nanoleaves, and they demonstratedhigh reactivity towards electrocatalytic reduction of oxygen in a 0.1 M NaOHelectrolyte: 3.0 increase in the specific activity and 2.7 increase in themass activity compared with a commercial Pt/C catalyst (at 0 V *vs.*Hg/HgO). The electrocatalytic activity enhancement can be attributed to theirunique nanoleaves structure, i.e. more Pd (111) facets, large surface area andmore resistance to Pd oxide formation.

See more of this Session: Nanomaterials for Hydrogen Production and Fuel Cells II
See more of this Group/Topical: Topical 5: Nanomaterials for Energy Applications

(319d) Molecular Optimization of Multiply-Functionalized Mesoporous Films with Ion Conduction Properties

Donghun Kim[1], George L. Athens[1], Yu Seung Kim[2] and Bradley F. Chmelka[1],
(1)Chemical Engineering, University of California, Santa Barbara, Santa Barbara, CA,
(2)Los Alamos National Laboratory, Los Alamos, NM

Theperformance and stability of hydrogen fuel cells could be improved by operationat elevated temperatures (>100), which would provide faster kinetics ofthe anode and cathode reactions, higher permissible CO levels in the hydrogenfeed stream, more effective water management, and efficient heat recovery.However, currentindustrial proton exchange membranes based on perfluorinated sulfonic acids (PFSAs, e.g., Nafion^{TM})limit hydrogenfuelcelloperationto relativelylow temperatures (<100) and high humidities, which lead otherwise to reducedproton conductivities as the hydrophilic ion-conducting channels dehydrate and becomedisconnected. We have developed anovel proton-exchange-membrane material based on PFSA-functionalizednanostructured aluminosilica for the operation of hydrogen fuel cells atelevated temperatures (>100) and in dry environments. Theseorganic-inorganic hybrid membranes are designed to maintain theirproton-conducting channels by exploiting their robust aluminosilica frameworkswith interconnected pores that remain open, independent of the extent ofhydration, thus sustaining proton-conduction properties at elevatedtemperatures and low humidities. The mesoporous membranes are synthesized by using blockcopolymers to direct theformation of cubic mesostructured silica films, which are subsequently functionalizedwith aluminosilica and PFSA species. Molecular, mesoscopic, and macroscopicproperties of the multiply-functionalized films have been characterized andcorrelated at each stage of the syntheses by NMR, small-angle X-ray and neutronscattering, transmission electron microscopy, elemental analysis, adsorption,and conductivity measurements. The resulting materials have a novel combinationof stable mesopores, high hydrophilicities, and maintain high protonconductivities at elevated temperatures and under low humidity conditions.

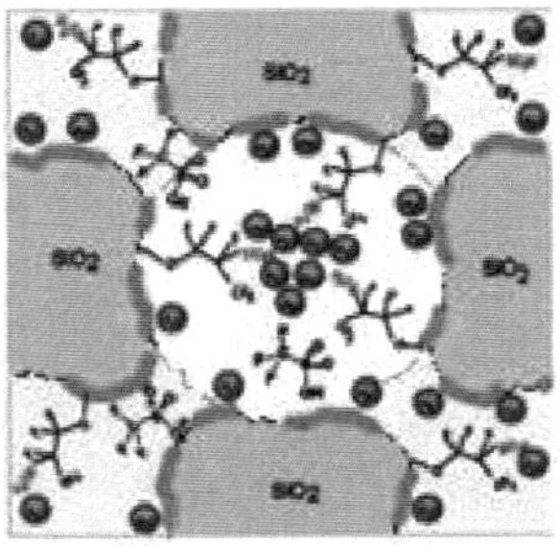

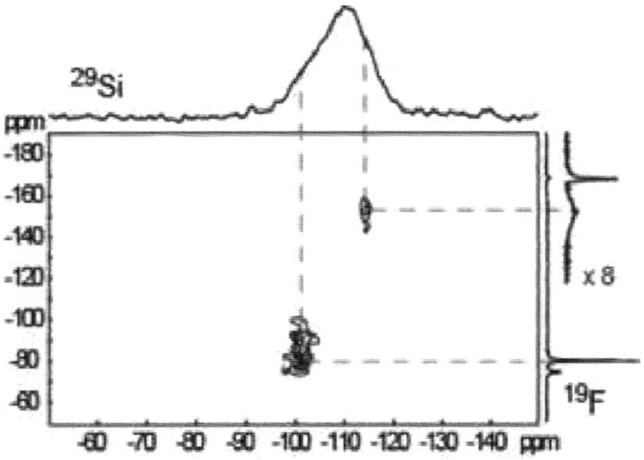

(319e) Morphology and Transport Properties of Sulfonated Poly(styrene-isobutylene-styrene) with Counter Ion Substitution

Sonia L. Aviles-Barreto and David Suleiman, Chemical Engineering, University of Puerto Rico, Mayaguez Campus, Mayaguez, PR

In this study, the transport properties of poly(styrene-isobutylene-styrene) (SIBS) were determined as function of sulfonation level and counter ion substitution for gas sensors and fuel cell applications. Sulfonated SIBS was neutralized with with +2 and +1 cations: Mg^{+2}, Ca^{+2}, Ba^{+2}, Mn^{+2}, Cu^{+2}, K^{+1}; to create highly selective polymer-metal nanocomposite membranes. Absorption limitations and their effect on the membrane transport were investigated for sulfonated and cross-linked membranes through methanol (MeOH) swelling and permeability experiments. Results show that MeOH permeability increases with sulfonation level until a maximum suggesting an optimum ion exchange capacity (IEC) level, while counter ion substitution decreases permeability. Morphologies studies using Small-angle X-ray scattering (SAXS) were performed to determine size and shape of nanostructured channels. Although increasing sulfonation levels influences morphology towards a more ordered structure (*i.e.*, spherical, hexagonal cylinders, lamellar), increasing sulfonation beyond a certain level creates an amourphous morphology perhaps with additional interconnections in the 3-D network. To complement the studies, selectivity (i.e., proton conductivity/methanol permeability) of sulfonated membranes was determined and compared to Nafion® 117.

See more of this Session: Nanomaterials for Hydrogen Production and Fuel Cells II
See more of this Group/Topical: Topical 5: Nanomaterials for Energy Applications

(401a) Photonic Metamaterials: Challenges and Oppurtunities

Costas M. Soukoulis, Ames Laboratory and Iowa State University, Ames, IA

In the last decade, a new area of photonics research has emerged, that has given the ability to produce materials with entirely novel electromagnetic properties. Known as metamaterials for their ability to take beyond conventional materials. Clearly, the field of metamaterials can develop mould-breaking technologies for a plethora of applications, where control over light (or more generally electromagnetic radiation) is a prominent ingredient—among them telecommunications, solar energy harvesting, biological and THz imaging and sensing, optical isolators and polarizers. In this talk, I give an introduction into this emerging field, review recent progress, and highlight remaining challenges and opportunities.

See more of this Session: Plasmonic Nanomaterials for Energy Applications
See more of this Group/Topical: Topical 5: Nanomaterials for Energy Applications

(401b) Trapping Light with Dark Plasmons In Nanoparticle Arrays

Teri W. Odom, Department of Chemistry and Materials Science and Engineering, Northwestern University, Evanston, IL

Metal nanostructures concentrate optical fields into highly confined, nanoscale volumes, which is important for applications such as plasmonic nanolasers and white-light generation. However, the short lifetimes of the measured or "bright" plasmons preclude amplification of local fields because of rapid depletion of the plasmon energy. To overcome this problem, we will describe a new type of "dark" or subradiant plasmon that can be supported by 2D arrays of large (> 100 nm, all three dimensions) gold particles. This dark lattice plasmon can be tuned over a broad range of wavelengths simply by changing the nanoparticle height. At resonance, strong coupling between out-of-plane nanoparticle dipolar moments suppresses radiative decay, trapping light in the plane of the array and strongly localizing optical fields on each nanoparticle. This new mechanism can open applications for subradiant plasmons because height-controlled nanoparticle arrays can be manufactured over wafer-scale areas on a variety of substrates.

See more of this Session: Plasmonic Nanomaterials for Energy Applications
See more of this Group/Topical: Topical 5: Nanomaterials for Energy Applications

(401c) Tailoring Thermal Emission Using Photonic and Plasmonic Metal Nanostructures

Prashant Nagpal, Physical Chemistry and Applied Spectroscopy, Los Alamos National Labratory, Los Alamos, NM

Radiative thermal emission from hot bodies is broad in energy and is non-directional. This poses problems for efficient energy conversion of this emitted radiation from hot body sources (eg. from sun) into electricity using an infrared photocell. Metal nanostructures, patterned as three-dimensional photonic crystals, selectively tailor the light-matter interaction near the photonic bandgap and the increased photonic density of states is used to selectively emit light of particular frequency from a hot metallic nanostructure. Here I will present results from selective tailoring of thermal emission at low temperatures for development of efficient thermophotovolatic emitters. This can possibly serve as an efficient route for harnessing solar-thermal energy and waste heat, and converting it into electricity with an upper conversion limit of over 60 percent.

Another possible route towards tailoring thermal emission is utilizing surface plasmons. While these hybrid photon-electron waves on metal surfaces are typically excited using light or fast electrons, recently we have explored excitation of surface plasmon polaritons using heat. I will discuss recent results on fabricating ultrasmooth patterned surface with nanometer scale precision for low loss propagation of surface plasmons. These patterned surfaces will then be combined with designed nanostructures for modulating the frequency of emitted radiation, and also the directionality of the emitted glow. Such unprecedented control over hot body radiation opens up new opportunities and challenges for utilizing plasmonics, while also opening up alternate route towards efficient photovoltaic conversion.

See more of this Session: Plasmonic Nanomaterials for Energy Applications
See more of this Group/Topical: Topical 5: Nanomaterials for Energy Applications

(401d) Self-Assembled Plasmonic Electrodes for High-Performance Organic Photovoltaic Cells

Wade A. Luhman[1], Si Hoon Lee[2], Timothy W. Johnson[2], Russell J. Holmes[1] and **Sang-Hyun Oh**[2], (1) Chemical Engineering and Materials Science, University of Minnesota, Minneapolis, MN, (2)Electrical and Computer Engineering, University of Minnesota, Minneapolis, MN

Thin Ag films perforated with subwavelength hole arrays are used as transparent conducting electrodes for organic photovoltaic cells. Plasmonic electrodes are fabricated using nanosphere lithography to create hexagonal nanohole arrays. Organic photovoltaic cells constructed using a nanopatterned Ag anode show power conversion efficiencies that exceed those of devices constructed on conventional indium-tin-oxide, independent of light polarization. In comparison to cells constructed on unpatterned Ag, the power conversion efficiency is noted to double with patterning.

See more of this Session: Plasmonic Nanomaterials for Energy Applications
See more of this Group/Topical: Topical 5: Nanomaterials for Energy Applications

(401e) Coupling Plasmons In Metal Nanoparticles to Excitons In Semiconductor Nanocrystals

Matthew Pelton, Center for Nanoscale Materials, Argonne National Laboratory, Argonne, IL

Coupling plasmon resonances in metal nanoparticles to excitons in semiconductor nanocrystals, or quantum dots, opens up new possibilities for light harvesting and management of optical energy at the nanoscale. These nanoparticles can be synthesized and organized using wet-chemical methods, providing a bottom-up method for the development of nanomaterials with new optical functionality. Exploiting plasmon-exciton interactions will require a fundamental understanding of the relationship between nanometer-scale structure of hybrid metal-semiconductor materials and their optical properties, particularly in the limit of strong coupling where new optical phenomena can arise. I will present progress in this direction, including calculations showing that a single quantum dot can induce transparency in a plasmonic nanoresonator, a strategy to overcome radiative losses by coupling to dark plasmon modes, and measurements on a model system consisting of single semiconductor nanocrystals near a metal surface.

See more of this Session: Plasmonic Nanomaterials for Energy Applications
See more of this Group/Topical: Topical 5: Nanomaterials for Energy Applications

(401f) Influence of Plasmonic Au Nanoparticles On the Photoactivity of Fe2O3 Electrodes for Water Splitting

Elijah Thimsen[1], Florian Le Formal[2], Michael Graetzel[2] and Scott C. Warren[2], (1)Material Science Division, Argonne National Laboratory, Argonne, IL, (2)Institute of Chemical Sciences and Engineering, Ecole Polytechnique Fdrale de Lausanne, Lausanne (VD), Switzerland

An experimental study of the influence of gold nanoparticles on α-Fe_2O_3 photoanodes for photoelectrochemical water splitting is described. A relative enhancement in the water splitting efficiency at photon frequencies corresponding to the plasmon resonance in gold was observed. This relative enhancement was observed only for electrode geometries with metal particles that were localized at the semiconductor-electrolyte interface, consistent with the observation that minority carrier transport to the electrolyte is the most significant impediment to achieving high efficiencies in this system.

See more of this Session: Plasmonic Nanomaterials for Energy Applications
See more of this Group/Topical: Topical 5: Nanomaterials for Energy Applications

(401g) Nanomanufacturing of Multicomponent Plasmonic Nanogels and Interfaces with Broadband Solar Absorption Capability

Tao Cong, Satvik Wani, Peter Paynter and Radhakrishna Sureshkumar, Department of Biomedical and Chemical Engineering, Syracuse University, Syracuse, NY

Thin film photovoltaics (PVs) offer much potential in cost-effective harvesting of solar energy through efficient use of the semiconductor materials, typically in the form of thin films of thickness 1-2 microns. However, the efficiency of single junction thin film PVs is significantly lower compared to their wafer-based counterparts due to poor light trapping by such devices. Improving the efficiency of thin-film PVs using plasmonic interfaces is an area of active research. A promising approach is the incorporation of a light trapping layer that consists of noble metal nanoparticles (NPs) onto the PV device [1-3]. Nanostructured plasmonic interfaces for this purpose have been fabricated by using lithography, vapor deposition, dewetting of thin metal films by ns and fs pulsed lasers and wet chemistry using self-assembled monolayers [4-13]. However, economical scale up and adaptation of such processes to fabricate interfaces with multiple species/shapes/sizes in a controllable and repeatable fashion are not straightforward.

In this work, we show how a nanostructured interface consisting of multiple metals and/or multiple shapes (e.g. sphere, rod) of a given metal can serve as a "broadband antenna" that would trap light in the UV-visible range. Specifically, a network of wormlike surfactant micelles (WLMs) in an aqueous solution is used as a template for producing stable multicomponent suspensions of Au and/or Ag NPs with desired optical properties. Such suspensions, hereafter referred to as plasmonic nanogels (PNGs), exhibit a long shelf life (~ weeks) and remarkable color uniformity. The structure of the PNGs was studied by small angle X-ray scattering, cryogenic transmission electron microscopy (Cryo-TEM) and rheological experiments. These studies, together with Molecular Dynamics simulations, support a mechanism of self-assembly in which the nanoparticles bridge the micelle fragments to form a stable double network. As evidenced by UV-visible transmission spectroscopy, the shape, size and concentration of the NPs can be varied to tune the optical properties of the PNGs in a way that they absorb radiation over a broadband of wavelengths. Multicomponent PNGs reported in this work have relatively low viscosity and low elastic modulus. Hence they are processable by conventional coating techniques. We employed spin- and dip-coating to produce multicomponent plasmonic interfaces. The structure and optical properties of such interfaces and their integration into PV devices will be discussed.

Acknowledgements: We acknowledge National Science Foundation grant CBET-1049454 for partial support of this research. *Syracuse University has filed a provisional patent application based on the findings of this work.*

References

1. H. A. Atwater and A. Polman, Nat. Mater. **9** (3), 205 (2010).

2. S. Pillai, K. R. Catchpole, T. Trupke, and M. A. Green, J. Appl. Phys. **101**, 093105 (2007).

3. J. Trice, H. Garcia, R. Sureshkumar, and R. Kalyanaraman, Proc. SPIE **6648**, 66480L (2007).

4. E. M. Hicks, S. Zou, G. C. Schatz, K. G. Spears, R. P. Van Duyne, L. Gunnarsson, T. Rindzevicius, B. Kasemo, and M. Käll, Nano Lett. **5** (6), 1065 (2005).

5. L. M. Campos, I. Meinel, R. G. Guino, M. Schierhorn, N. Gupta, G. D. Stucky, and C. J. Hawker, Adv. Mater. **20** (19), 3728 (2008).

6. Y. H. Lanyon, G. De Marzi, Y. E. Watson, A. J. Quinn, J. P. Gleeson, G. Redmond, and D. W. M. Arrigan, Anal. Chem. **79** (8), 3048 (2007).

7. Y. Wu, J. D. Fowlkes, P. D. Rack, J. A. Diez, and L. Kondic, Langmuir, 739 (2010).

8. L. Kondic, J. A. Diez, P. D. Rack, Y. Guan, and J. D. Fowlkes, Phys. Rev. E **79** (2), 026302 (2009).

9. C. Favazza, J. Trice, R. Kalyanaraman, and R. Sureshkumar, Appl. Phys. Lett. **91**, 043105 (2007).

10. J. Trice, C. Favazza, D. Thomas, H. Garcia, R. Kalyanaraman, and R. Sureshkumar, Phys. Rev. Lett. **101** (1), 17802 (2008).

11. C. Favazza, R. Kalyanaraman, and R. Sureshkumar, Nanotechnology **17**, 4229 (2006).

12. A. I. Kuznetsov, J. Koch, and B. N. Chichkov, Appl. Phys. Mater. Sci. Process. **94** (2), 221 (2009).

13. L. A. Porter Jr, H. C. Choi, J. M. Schmeltzer, A. E. Ribbe, L. C. C. Elliott, and J. M. Buriak, Nano Lett. **2** (12), 1369 (2002).

See more of this Session: Plasmonic Nanomaterials for Energy Applications
See more of this Group/Topical: Topical 5: Nanomaterials for Energy Applications

(470a) Silicon-Based Bulk Nanostructured Thermoelectric Generators

Akram Boukai, Materials Science and Engineering, University of Michigan, Ann Arbor, MI

State-of-the-art thermoelectric generators are typically based on exotic alloys and have low device efficiencies. Therefore, research targeting new discoveries in earth abundant thermoelectric materials that increase device efficiency are needed. This talk will present recent progress on bulk nanostructured thermoelectric materials that are silicon-based. These include functional silicon nanowire devices and silicide/silicon eutectics. These bulk nanostructured thermoelectric devices exploit dramatic reductions of thermal conductivity due to phonon scattering at disordered interfaces, which is expected to increase ZT. Thermoelectric properties, structural characterization, and device efficiencies will be presented.

See more of this Session: Nanomaterials for Thermal-to-Electric Conversion I
See more of this Group/Topical: Topical 5: Nanomaterials for Energy Applications

(470b) Advanced Nanostructures for Thermoelectric Energy Harvesting

Yue Wu, School of Chemical Engineering, Purdue University, West Lafayette, IN

Substantial efforts have been devoted to use nanostructured materials for thermoelectric energy harvesting and solid-state cooling in the past decade. However, the majority of the research is still limited in lab scale due to the incapability to mass-produce well-defined nanostructured materials with low yet industrial-compatible process. In addition, a lot of widely used thermoelectric materials contain toxic and expensive elements that prevent the large-scale deployment of the thermoelectric devices. In this presentation, we will discuss our research on the development of mass production of molecular scale nanostructures of chalcogenides and metal oxides, as well as their heterostructures, for the manufacture of thermoelectric generators operating at different temperature ranges. Particularly, we will talk about our approach to discover and investigate the non-toxic and abundant nanostructured materials to achieve an environmentally friendly process. Our preliminary research indicated that thermoelectric figure of merit (ZT) close to 2 can be achieved in the molecular scale nanowires of certain chalcogenide due to significantly enhanced power factor and reduced thermal conductivity as a result of quantum confinement.

See more of this Session: Nanomaterials for Thermal-to-Electric Conversion I
See more of this Group/Topical: Topical 5: Nanomaterials for Energy Applications

(470c) Molecular and Hybrid Solution Processible Thermoelectrics

Rachel A. Segalman[1], Shannon Yee[2], Boris Russ[2], Nelson Coates[2] and Jeffrey Urban[3], (1)Department of Chemical and Biomolecular Engineering, University of California, Berkeley, Berkeley, CA, (2)University of California at Berkeley, Berkeley, CA, (3)Lawrence Berkeley National Laboratories, Berkeley, CA

Thermoelectric materials for energy generation have severaladvantages over conventional power cycles including lack of moving parts,silent operation, miniaturizability, and CO2 freeconversion of heat to electricity. Excellentthermoelectric efficiency requires a combination of high thermopower(S, V/K), high electrical conductivity (σ, S/cm), and low thermalconductivity (κ, W/mK). To date the bestmaterials available have been inorganic compounds with relatively low earthabundance and highly complex, vacuum processing routes (and hence greaterexpense), such as Bi_2Te_3. Molecular materials andhybrid organic-inorganics bring the promise ofinexpensive, solution processible, mechanicallydurable devices. While highlyconductive polymers are now common place, theygenerally demonstrate low thermopower. Our work onmolecular scale junctions demonstrates that nanostructuringof organics allows them to act as thermionic filters between inorganicjunctions which can lead to enhanced thermoelectric properties. We have takeninspiration from this fundamental understanding to design material systems inwhich we combine a high electrical conductivity, low thermal conductivitypolymer with a nanoparticle that contributes high thermopower. Additionally, the work functions of the two materials are well-aligned which introduces the possibility of thermionicfiltering at the interface and an additional boost to the power factor. The combination of these effects results in a new hybrid,solution processible material with a thermoelectricfigure of merit within an order of magnitude of the Bi_2Te_3. In this talk, I will discuss both the use ofthermoelectric measurements to gain insight to molecular junctions and how thisinsight translates to design principles for polymer and hybrid thermoelectrics.

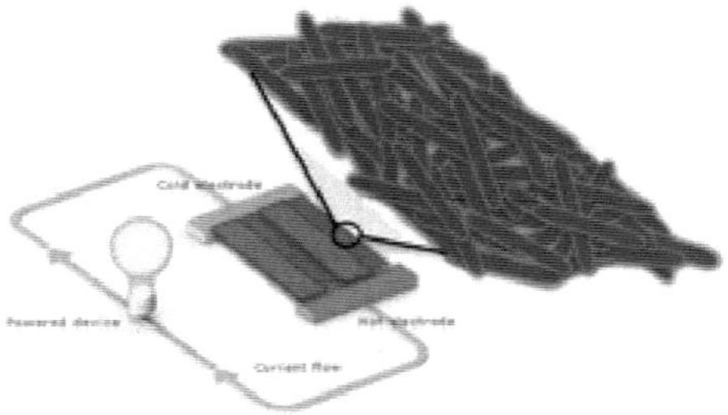

See more of this Session: Nanomaterials for Thermal-to-Electric Conversion I
See more of this Group/Topical: Topical 5: Nanomaterials for Energy Applications

(532a) Ultra-Rapid Elimination of Biofilms On Metal and Ceramic Surfaces Via the Combustion of a Nanoenergetic Coating

Shramik Sengupta[1], Rajagopalan Thiruvengadathan[2], Byung-Doo Lee[1], Brandon M. Smith[1], Sachidevi Puttaswamy[1], Keshab Gangopadhyay[3] and Shubhra Gangopadhyay[3], (1)Biological Engineering, University of Missouri, Columbia, MO, (2)University of Missouri Columbia, Columbia, MO, (3) Electrical and Computer Engineering, University of Missouri, Columbia, MO

In Heat Exchangers and Cooling-Water Systems, which are an integral part of a wide variety of industrial processes, the presence of biofilms can reduce the effective heat transfer coefficient by as much as 50%. Biofilms consist of microbials colonies embedded on the surface of the equipment, and protected by a layer of extra-cellular polymeric substances secreted by the same microbes.

Removal of the biofilms requires that the equipment be taken "offline" and subjected to chemical and physical treatments that not only take a long time, but also require the use of toxic chemicals. Reducing the down-time for the equipment and the development of processes that do not require the use and disposal of toxic chemicals, is thus highly desirable from both economic and environmental points of view.

We present here a novel process that uses nano-energetic materials to eliminate biofilms in just a few seconds. The process involves spray-coating of an approximately 100 μm thick layer of nano-energetic material on top of the biofilm, allowing it to dry, and igniting the dried coating to incinerate the biofilm. The nanoenergetic material used is a two component mixture consisting of aluminum nanoparticles (fuel) dispersed in a THV220A (fluoropolymer oxidizer) matrix. Upon ignition, the Al nanoparticles react with THV 220A exothermically, producing high temperatures (> 2500K) for an extremely brief period (~100mS) that destroys the biofilm underneath, along with the bacteria harbored within. However, due to the small mass of nanoparticles present (~1 mg/cm^2 of coating), the total *amount* of heat produced is low (~0.1 KJ/cm^2) and the underlying metal or ceramic surface does not suffer observable damages. Since the byproducts of the process consist only of carbonaceous residue (that can be easily washed away) and gaseous combustion products, the process is not extremely fast, but also environmentally friendly.

See more of this Session: Green and Renewable Energy Aspects of Nanotechnology
See more of this Group/Topical: Topical C: Environmental Aspects, Applications, and Implications of Nanomaterials and Nanotechnology

(532b) Life Cycle Energy Analysis and Midpoint Assessment of Multimegawatt Wind Turbines with Polymer Nanocomposite Blade Material

Laura A. Merugula[1], Bhavik R. Bakshi[1] and Vikas Khanna[2], (1)Lowrie Department of Chemical and Biomolecular Engineering, The Ohio State University, Columbus, OH, (2)Department of Civil and Environmental Engineering, University of Pittsburgh, Pittsburgh, PA

Life cycle analysis ofvapor-grown carbon nanofiber (VGCNF or CNF) manufacturing shows production ishighly energy-intensive with a cumulative energy demand that is influenced byfeedstock selection and dominated by processing energy needed to maintainreactor temperature for the catalytic pyrolysis of the hydrocarbon feed.However, at relatively low mass percentages, CNFs can reinforce polymercomposites yielding significant improvements in properties such as mode-1delamination resistance, vibrational damping, tensile strength, fatigueresistance and Young's modulus. A new method termed "pre-binding" has beendeveloped to reinforce polymer composites with CNFs at the interface of longfibers and resin. This method improves the manufacturability of large pieceswith CNF-reinforced polymer nanocomposites (PNCs) by overcoming processingissues inherent to previous methods of PNC fabrication. Materials scientistsare considering use of this emerging material for wind turbine blade productionto meet the demand for larger, more resilient utility-scale wind power plants.

Wind energy converters (WECs)on-line today are a relatively new technology with installations of multimegawattratings having been introduced to the modern market only in the last 15-20years. The largest 3-blade horizontal axis wind turbine (HAWT) rotor size inoperation is 126 meters in diameter with generator ratings from 5 to 7 MW whilethe growing market is dominated by new installations from 1 to 3 MW. Apparentlimits of fiberglass technology have been reached as rotor blades beyond about40 meters generally require some inclusion of carbon fiber to address stiffnessand edgewise fatigue, which become critical at this length. The less densecarbon fiber also leads to a positive deviation in trends of weight toincreasing size. Recent press releases communicate progress in R&D formeeting industrial-governmental collaborative goals for 20 MW plants possiblyby 2020 using available technologies with rotors expected to have diametersnear 200 meters.

In an even shorter timespan, numerous nanotechnology-enabled products have been introduced to theconsumer market with over 1,300 manufacturer-identifiedproducts populating The Project on Emerging Nanotechnologies consumer productinventory today. These products include automobile body parts with carbon nanotubes(CNTs) for enabling online electrostatic painting and a personal watercraftusing nanomaterials for reinforcement of the polymer composite for alightweight, stronger material.

While life cycle energyconsumption of CNF-reinforced PNCs for use in automobile body panels has beenevaluated, similar trends were not expected for life cycle energy in windturbines due to differences in operational phase energy consumption. Also,previous assessments of PNCs have not considered the method of pre-binding,which uses acetone as a vessel for dispersion and distribution of the CNFs ontoglass or carbon fiber matting.

Preliminary studies thatassumed replacement of polymer composite blade materials on a 2 MW and 5 MWplant with CNF pre-bound PNCs indicated a range of parameterization valuesunder which the energy return on investment (EROI) would not decrease. Thesestudies have been updated in a process-based life cycle assessment (LCA) toinclude more relevant process information and have been extended to a midpointanalysis using the CML and Traci methods to determine areas of focus foroptimization. An energy analysis was extended to a 5 MW WEC using cumulativeenergy demand and power generation figures from published LCA literature.

Impacts are reported perkilowatt-hour generation of electricity by the wind power plant and compared tothe base case, which is a 2 MW WEC in Middelgrund that is inventoried in the proprietaryEcoinvent database. In every feasible scenario, the EROI of wind power plants withthe PNC blade material remained competitive to thermoelectric power generation,and life cycle water consumption remained favorable. Where CNF production wasassumed to be based on benzene feedstock, energy analysis results were muchmore favorable than methane. Midpoint analysis, however, was assessed with methaneas the preferred hydrocarbon source for pyrolysis in nanofiber manufacture. Similartrends emerged under extension to the 5 MW system energy analysis.

Assuming a masscontribution of CNFs that ranges from 1 to 5 % by weight of each blade, theoverall mass of the nanomaterials is approximately 0.01-0.06% of the overallmass of the WEC system, yet there is direct correlation between the massfraction of the nanomaterials to nontrivial impacts per kWh compared to thebase case. Where studies have shown manufacturing of virgin nanomaterials tobe solvent-intensive, this study finds that the use phase of the nanomaterial,where CNFs are integrated into a product system, is highly solvent-intensive aswell. Current laboratory scale ratios of solvent spent to nanofibers dispersedare on the order of $5x10^2$ g solvent per g CNF. Maturation of thismaterial development must include a reduction of solvent use by optimization ofthe dispersion process and evaluation of solvent recovery systems for reusethat considers effects of nanomaterials in the operation. Photochemicaloxidation is a concern in terms of acetone emissions, the degree to which willbe dependent on recovery, reuse, and emissions controls. Also, the degree ofsolvent fresh feed required has an impact on the overall system's globalwarming potential and eutrophication potential due to the manufacturing phaseof the acetone.

Evaluating impacts of emergingtechnologies is complicated by the fast-pace of development and deployment,significant gaps in data and uncertainties about the future. Compilation oflife cycle data is unwieldy at best due to complex supply chains and regionalvariations in categories as energy supply and transportation. This is furthercomplicated by incomplete understanding of ecosystem service impacts, thoughwork is underway to quantify and compile ecosystem services for inclusion in LCA.System performance improvements will not occur with PNC integration inisolation of other technological changes within WEC design and operation, yet analysesas this are meant to give insight to areas of concern by identifying impactsand order of magnitude effects associated with this emerging material for theassumed use.

See more of this Session: Green and Renewable Energy Aspects of Nanotechnology
See more of this Group/Topical: Topical C: Environmental Aspects, Applications, and Implications of Nanomaterials and Nanotechnology

(532c) Advanced Oxidation Processes with Carbon Nanotubes: Surface-Promoted Formation of Hydroxyl Radical During Ozonation

Rebekah L. Oulton[1], Michael J. Nalbandian[1], Howard A. Fairbrother[2], Kevin Wepasnick[2] and David M. Cwiertny[1], (1)Chemical & Environmental Engineering, University of California, Riverside, Riverside, CA, (2)Department of Chemistry, Johns Hopkins University, Baltimore, MD

Advanced Oxidation Processes (AOPs), which utilize hydroxyl radical (•OH) to chemically degrade contaminants, are becoming increasingly utilized in water and wastewater treatment for disinfection and removal of organic micropollutants via chemical oxidation reactions. Catalytic ozonation represents an alternative AOP in which a solid substrate such as activated carbon or common metal oxides are used to promote •OH formation from ozone. This project explores the efficacy of using multi-walled carbon nanotubes (MWCNTs) as new materials for promoting •OH production during ozonation. Using para-chlorobenzoic acid (*p*-CBA) as a hydroxyl radical probe, aqueous suspensions of functionalized (i.e., surface oxidized) and nonfunctionalized MWCNTs increased •OH concentrations in bulk solution by an order of magnitude or more over levels observed with ozone alone and resulted in •OH levels comparable to more traditional ozone-based AOPs (e.g., hydrogen peroxide with ozone). Production of •OH was greatest on oxidized MWCNTs, and evidence suggests this enhanced •OH production cannot be explained solely by the increased dispersivity of oxidized MWCNTs in aqueous suspensions. Chemical routes for MWCNT functionalization were used to systematically vary the surface oxygen content of MWCNTs, and the ability of these functionalized materials to generate •OH was subsequently determined in batch ozonation experiments, again using *p*-CBA as a radical probe. Complementary surface analysis with X-ray photoelectron spectroscopy (XPS) combined with chemical derivatization was used to correlate •OH production not only to total MWCNT surface oxygen content, but also to the concentration of specific functional groups (hydroxyl, carbonyl and carboxylic acid groups) resulting from oxidation. Results to date show that •OH production increases with MWCNT surface oxidation, and that carboxyl surface groups appear largely responsible for this increase in •OH production.

See more of this Session: Green and Renewable Energy Aspects of Nanotechnology
See more of this Group/Topical: Topical C: Environmental Aspects, Applications, and Implications of Nanomaterials and Nanotechnology

(532d) Evaluating the Dynamics of An Integrated Electrokinetic and Zero-Valent Iron Nanoparticle System for Treatment of Hexavalent Chromium In Groundwater

Ryan Thacher and Massoud Pirbazari, Sonny Astani Department of Civil and Environmental Engineering, University of Southern California, Los Angeles, CA

Electrokinetic technology is a promising groundwater treatment strategy capable of transporting ionic contaminants through groundwater by application of an electric potential across a contaminated area. While this is effective in the transport and concentration of contaminants at an electrode, complimentary remedial action may still be required to enhance treatment of the contaminants. Zero-valent iron nanoparticles (nZVI) may potentially offer an ideal solution for this genre of *in-situ* environmental remediation. Their small size coupled with their high surface area to volume ratio as well as reductive capacity, make nZVI a seemingly ideal medium for subsurface remediation where other classical remediation strategies such as excavation are not possible. This may be in deep aquifers, urban areas, or under existing infrastructure.

A comprehensive understanding of the fate and transport of nZVI in the subsurface is just as critical to *in-situ* remediation as the high reactivity. Upon introduction to an aquifer environment, nZVI must be readily dispersible and remain stable so that they reach the targeted contamination zone. Several researchers have investigated transport of nZVI through porous media, however most of the published work is conducted in idealized batch studies. These investigations have indeed provided a good knowledge base, but limited information exists concerning the fate, transport, and toxicity of nZVI in the natural environment. The challenging issues at hand include: nanoparticle agglomeration, nZVI storage and age, components of soil matrices, aquifer hydraulic properties, presence of natural organic matter, pH, and redox potential, among others.

This study investigates the treatment of simulated groundwater containing hexavalent chromium (Cr^{VI}), a known carcinogen, in a dynamic system using combined electrokinetic and zero-valent iron nanoparticle technology. Continuous flow columns are charged with simulated natural soils, followed by saturation with solutions containing Cr^{VI}, variable humic acid concentration, and at multiple pH values. Upon saturation, electric potential is applied, and dilute solutions of nZVI are added at regular intervals for the entirety of the experiment.

This presentation will discuss the influence of natural aquifer characteristics such as pH, natural buffer, and natural organic matter content on the transport of nZVI through subsurface environments, and how this impacts treatment of contaminants. Furthermore, this research will evaluate the dynamics of coupling electrokinetic and nZVI technologies in decontamination of the subsurface environment.

See more of this Session: Green and Renewable Energy Aspects of Nanotechnology
See more of this Group/Topical: Topical C: Environmental Aspects, Applications, and Implications of Nanomaterials and Nanotechnology

(532e) Biodegradation of Carbazole by Microbial Cells Coated with Magnetite Nanoparticles

Jianmin Xing, National Key Lab of Biochemical Engineering, Institute of Process Engineering, Chinese Academy of Sciences, Beijing, China

Microbes can remove nitrogen from carbazole through a series of enzyme-catalyzed reactions. Microbial cells of *Klebsiella* sp. LSSE-H2 were coated with magnetic Fe_3O_4 nanoparticles, which were synthesized by a coprecipitation method followed by modification with ammonium oleate. The average size of the magnetic particles was found to be in the range from 10 to 15 nm. Transmission electron microscopy (TEM) cross section analysis of the cells showed further that the Fe_3O_4 nanoparticles were strongly absorbed by the surfaces of the cells and coated the cells. The coated cells had superparamagnetic properties. The coated cells not only had the same activity as free cells but could also be reused. Biodenitrogenation rate was increased with the addition of Tween 80. The activity decreased less than 23% after three recycles.

See more of this Session: Green and Renewable Energy Aspects of Nanotechnology
See more of this Group/Topical: Topical C: Environmental Aspects, Applications, and Implications of Nanomaterials and Nanotechnology

(542a) Formation and Control of Thermoelectric Composites At the Nanoscale

G. Jeffrey Snyder, Caltech, Pasadena, CA

Many new high efficiency thermoelectric materials are multi-phase composites at the nanometer scale. These include materials such as PbTe alloys, and recently Zn4Sb3, which contain small ~10nm nanoparticles that can be difficult to control. Typically the goal is to reduce the lattice thermal conductivity of the matrix by scattering phonons at the interfaces of the particles. Thus it is not the small size of the particles but the spacing between particles that matter. We have found that in PbTe with large (~50nm – 200nm) nanoparticles of Ag_2Te high $zT \sim 1.5$ can be obtained through the independent control of thermal conductivity reducing Ag_2Te large nanoparticles and electronic doping with La.

In general, because electrons are also scattered, a net improvement in thermoelectric efficiency will depend strongly on the spacing, size, shape, orientation and interfacial properties of the particles. At Caltech we have extensively studied the formation and control of composite microstructure in PbTe-Sb_2Te_3, identifying three distinct mechanisms for forming composites and controlling the characteristic length scales down to the nanometer scale. The coherent interfaces of these composites may help retain high carrier mobility.

[1] G. J. Snyder, E. S. Toberer. *Nature Materials* **7**, p 105 - 114 (2008).

[2] D.L. Medlin and G.J. Snyder "Interfaces in Bulk Thermoelectric Materials" *Current Opinion in Colloid & Interface Science* (2009)

[3] Yanzhong Pei, Jessica Lensch-Falk, Eric S. Toberer, Douglas L. Medlin, and G. Jeffrey Snyder “High Thermoelectric Performance in PbTe due to Large Nanoscale Ag_2Te Precipitates and La Doping” *Advanced Functional Materials* **21**, 241 (2011)

See more of this Session: Nanomaterials for Thermal-to-Electric Conversion II: Thermoelectrics, Thermophotovoltaics and Thermopower
See more of this Group/Topical: Topical 5: Nanomaterials for Energy Applications

(542b) New Concepts In Energy Transport within Carbon Nanotubes: Thermopower Waves

Michael Strano, Chemical Engineering, Massachusetts Institute of Technology, Cambridge, MA

Our laboratory has been interested in how carbon nanotubes (CNTs) can be utilized to illustrate new concepts in molecular and energy transfer. We predict and demonstrate the concept of thermopower waves for energy generation. Coupling an exothermic chemical reaction with a thermally conductive CNT creates a self-propagating reaction wave driven along its length. We realize such waves in multi-walled CNTs and show that they produce concomitant electrical pulses of high specific power-- greater than 7 kW/kg. Such waves of high power density may find uses as unique energy sources.

Choi W, Hong S, Abrahamson JT, Han JH, Song C, Nair N, Baik S, Strano MS: Chemically driven carbon-nanotube-guided thermopower waves. NATURE MATERIALS, 9 (2010) 423-429.

See more of this Session: Nanomaterials for Thermal-to-Electric Conversion II: Thermoelectrics, Thermophotovoltaics and Thermopower
See more of this Group/Topical: Topical 5: Nanomaterials for Energy Applications

(542c) Thermal Plasmonics As a Route to Photovoltaics?

David J. Norris, Chemical Engineering and Materials Science, University of Minnesota, Minneapolis, MN

Patterned metallic films allow the generation and manipulation of special electromagnetic waves known as surface plasmons that propagate along a metal interface. Because these waves allow the concentration of light in nanometer-scale volumes, they have implications for fundamental phenomena as well as applications such as imaging, sensing, and solar cells. Thus, the field of plasmonics has arisen to study and utilize surface plasmons. While light is typically used to create these waves, in some cases it would be more convenient to have a non-optical source for surface plasmons. One route is thermal excitation (*i. e.*, heat). Here, we will explore hot plasmonic structures for obtaining new optical behavior. For example, we will examine metallic films patterned with a series of circular concentric grooves (a bull's eye pattern). We show that, when heated, these films can emit light that is amazingly narrow, both in terms of its spectrum and its angular divergence. Thus, a simple metallic foil can generate a highly directional beam of monochromatic light by a thermal process. This effect has implications for creating efficient thermophotovoltaic devices, which convert heat into electricity. During these studies, we also developed a simple, high-throughput method for obtaining ultrasmooth patterned metal films. Previously, roughness in such films has hindered the field of plasmonics. Therefore, our approach solves a critical problem and allows many high-quality plasmonic structures to be available for the first time.

See more of this Session: Nanomaterials for Thermal-to-Electric Conversion II: Thermoelectrics, Thermophotovoltaics and Thermopower
See more of this Group/Topical: Topical 5: Nanomaterials for Energy Applications

(542d) New Absorber and Emitter Designs and High Temperature Materials for Thermophotovoltaics

Paul V. Braun, Department of Materials Science and Engineering, University of Illinois at Urbana-Champaign, Urbana, IL

Thermophotovoltaic (TPV) solar cells, where broad-band solar radiation is absorbed by an intermediate, which then emits narrow-band thermal radiation towards a solar cell have been proposed for some time as a route for high efficiency solar energy harvesting. However, due to limitations in both design of the active components, as well as the high temperature stability of the materials making up these systems, the theoretical maximum efficiency of 85% for such devices has not been realized. We are currently exploiting emerging opportunities in the design and fabrication of metallic photonic crystals for TPV applications. We have now demonstrated tungsten based 3D photonic crystals which retain micron-scale structure even at temperatures of 1300 K, and are exploring other materials systems that may have even higher temperature stability. Through simulation and modeling, we have designed a number of absorber/emitter structures that match both the broad-band solar spectrum and the desired narrow band emission required for efficient single-junction solar cells. The self and directed assembly methods used to fabricate the TPV structures can produce complex large-area structures at low cost.

See more of this Session: Nanomaterials for Thermal-to-Electric Conversion II: Thermoelectrics, Thermophotovoltaics and Thermopower
See more of this Group/Topical: Topical 5: Nanomaterials for Energy Applications

(542e) Hybrid Nanowire/Polymer Composite for Thermoelectric Energy Harvest

Yue Wu, School of Chemical Engineering, Purdue University, West Lafayette, IN

Successfully decreasing the thermal conductivity while maintaining or increasing the power factor is the key to improve the efficiency for thermoelectric energy harvesting. Up to date, most research efforts in the development of thermoelectric materials have been devoted to use the rigid inorganic materials and little progress has been achieved to use the hybrid composite. In this presentation, we will discuss our research on the hybrid nanocomposite between inorganic thermoelectric nanowires and conductive polymers and our results indicated that significantly lowered thermal conductivity using the organic polymer matrix and reasonably good electrical conductivity and Seebeck coefficient could be achieved depending on the mixing ratio between nanowires and conductive polymer, which demonstrates a great potential to form hybrid thermoelectric devices with high flexibility as portable energy source.

See more of this Session: Nanomaterials for Thermal-to-Electric Conversion II: Thermoelectrics, Thermophotovoltaics and Thermopower
See more of this Group/Topical: Topical 5: Nanomaterials for Energy Applications

(609a) Exciton Engineering with Carbon Nanotubes and Graphene for Solar Energy Conversion: From Exciton Antennae to Nano-Heterojunctions

Michael Strano[1], Geraldine LC Paulus[2], Moon-Ho Ham[2], Chang Young Lee[2], Changsik Song[2], Kourosh Kalantar-Zadeh[3], Wonjoon Choi[4], Jae-Hee Han[1], Ryuichiro Maruyama[5], Esther S. Jeng[6], Daniel A. Heller[2], Woo-Jae Kim[1], Paul W. Barone[7] and Cristiano Fantini[8], (1)Chemical Engineering, Massachusetts Institute of Technology, Cambridge, MA, (2)Department of Chemical Engineering, Massachusetts Institute of Technology, Cambridge, MA, (3)School of Electrical and Computer Engineering, RMIT University, Melbourne, Australia, (4)Department of Mechanical Engineering, Massachusetts Institute of Technology, Cambridge, MA, (5)Materials Laboratories, Sony Corporation, Kanagawa, Japan, (6)Department of Chemistry, University of Illinois at Urbana-Champaign, Urbana, IL, (7)Chemical Engineereing, Massachusetts Institute of Technology, Cambridge, MA, (8)Departamento de Fisica, Universidade Federal de Minas Gerais, Belo Horizonte-MG, Brazil

Exciton engineering with carbon nanotubes and graphene for solar energy conversion: from exciton antennae to nano-heterojunctions

Michael S. Strano

Charles and Hilda Roddey Associate Professor of Chemical Engineering

66-566 Department of Chemical Engineering

77 Massachusetts Avenue

Cambridge, MA 02139-4307

Email: strano@MIT.EDU

phone: (617) 324-4323

fax: (617) 258-8224

http://web.mit.edu/stranogroup/

Our laboratory has been interested in how low dimensional materials, such as single walled carbon nanotubes or graphene, can augment and enhance solar conversion efficiencies and demonstrate new photonic concepts. We will discuss two recent examples of our work in this space.

In the first example, there is significant interest in combining carbon nanotubes with semiconducting polymers for photovoltaic applications because of potential advantages from smaller exciton transport lengths and enhanced charge separation. However, to date, bulk heterojunction (BHJ) devices have demonstrated relatively poor efficiencies, and little is understood about the polymer/nanotube junction. To investigate this interface, we fabricate a planar nano-heterojunction comprising well-isolated millimeter-long single-walled carbon nanotubes underneath a poly(3-hexylthiophene) (P3HT) layer (1). The resulting junctions display photovoltaic efficiencies per nanotube ranging from 3% to 3.82%, which exceed those of polymer/nanotube BHJs by a factor of 50-100. The increase is attributed to the absence of aggregate formation in this planar device geometry. It is shown that the polymer/nanotube interface itself is responsible for exciton dissociation. Typical open-circuit voltages are near 0.5 V with fill factors of 0.25-0.3, which are largely invariant with the number of nanotubes per device and P3HT thickness. A maximum efficiency is obtained for a 60 nm-thick P3HT layer, which is predicted by a Monte Carlo simulation that takes into account exciton generation, transport, recombination, and dissociation. We combine for the first time both optical T-matrix and kinetic Monte Carlo models to investigate the photocurrent generation in two state-of-the-art PHJ photovoltaics. The combined model takes into account the rates of exciton generation, transport, recombination and dissociation using literature values. By including the optical, electronic and structural properties of the different materials, we are able to predict the short-circuit current of recently reported P3HT/SWNT PHJ and also P3HT/PCBM PHJ solar cells from the literature. The experimental data for each of these devices show a maximum photocurrent output at a P3HT thickness of 60-65 nm, in contradiction to the expected value equal to the diffusion length of excitons in P3HT (8.5nm). The model demonstrates how a bulk exciton sink can explain this shifted maximum in the P3HT/SWNT case, whereas the maximum is mainly determined by PCBM interdiffusing in P3HT in the P3HT/PCBM case. This platform is promising for further understanding the potential role of polymer/nanotube interfaces for photovoltaic applications.

In the second example, there has been renewed interest in solar concentrators and optical antennas for improvements in photovoltaic energy harvesting and new optoelectronic devices. We dielectrophoretically assemble single-walled carbon nanotubes (SWNTs) of homogeneous composition into aligned filaments that can exchange excitation energy, concentrating it to the centre of core–shell structures with radial gradients in the optical bandgap (2). We find an unusually sharp, reversible decay in photoemission that occurs as such filaments are cycled from ambient temperature to only 357 K, attributed to the strongly temperature-dependent second-orderAuger process. Core–shell structures consisting of annular shells of mostly (6, 5) SWNTs (E_g = 1.21 eV) and cores with bandgaps smaller than those of the shell (E_g = 1.17 eV (7,5)–0.98 eV (8,7)) demonstrate the concentration concept: broadband absorption in the ultraviolet–near-infrared wavelength regime provides quasi-singular photoemission at the (8, 7) SWNTs. This approach demonstrates the potential of specifically designed collections of nanotubes to manipulate and concentrate excitons in unique ways.

1. Ham MH, Paulus GLC, Lee CY, Song C, Kalantar-zadeh K, Choi W, Han JH and **Strano MS**: Evidence for High-Efficiency Exciton Dissociation at Polymer/Single-Walled Carbon Nanotube Interfaces in Planar Nano-heterojunction Photovoltaics. ACS NANO, 4 (2010) 6251-6259

2. Han JH, Paulus GLC, Maruyama R, Jeng ES, Heller DA, Kim WJ, Barone PW, Lee CY, Choi JH, Ham MH, Song C, Fantini C, **Strano MS**: Exciton Antennas and Concentrators from Core-Shell and Corrugated Carbon Nanotube Filaments of Homogeneous Composition. NATURE MATERIALS, 9, 833 - 839 (2010).

See more of this Session: Semiconductor Nanowires and Nanotubes for Next Generation Photovoltaics From Fundamentals to Devices
See more of this Group/Topical: Topical 5: Nanomaterials for Energy Applications

(609b) Semiconductor Nanowire Fabric As a New Photovoltaic Material

Chet Steinhagen[1], Vahid Akhavan[1], Vince Holmberg[1] and **Brian A. Korgel**[2], (1)Chemical Engineering, University of Texas at Austin, Austin, TX, (2)Department of Chemical Engineering, University of Texas at Austin, Austin, TX

Significant quantities of semiconductor nanowires can be synthesized using solution-based methods, such as supercritical fluid-liquid-solid (SFLS) and solution-liquid-solid (SLS) growth. These nanowires are mechanically flexible because of their narrow (<50 nm) diameter and high aspect ratio, with lengths greater than millimeters in some cases. Because of the absence of crystallographic defects like grain boundaries, the nanowires are extremely strong, exhibiting nearly ideal mechanical strength. The nanowires can be formed into non-woven fabric, or paper, that is flexible and mechanically robust. We are in the process of trying to develop these nanowire fabrics as a new class of flexible photovoltaic material. This presentation will highlight the nanowire fabric itself--its properties and how it is made--as well as data from prototype photovoltaic silicon (Si) and CuInSe2 (CIS) nanowire fabric.

See more of this Session: Semiconductor Nanowires and Nanotubes for Next Generation Photovoltaics From Fundamentals to Devices
See more of this Group/Topical: Topical 5: Nanomaterials for Energy Applications

(609c) Solar Cells Based On Sensitized Nanowire Arrays

Jason B. Baxter, Department of Chemical and Biological Engineering, Drexel University, Philadelphia, PA

Solar cells can provide an abundant, clean, and sustainable source of electricity, but high costs have limited their implementation. The use of sensitized nanostructured architectures may enable both low-cost processing and high efficiency by decoupling the functions of light harvesting and charge transport into different materials. Sensitized solar cells consist of a bicontinuous interpenetrating network of electron- and hole-transporting materials with an interfacial absorber layer. In this architecture, interfacial recombination is the dominant loss process, so controlling the interfacial chemistry during deposition is critical. I will present our group's work on the use of ZnO nanowire arrays sensitized by thin CdSe coatings to produce extremely thin absorber (ETA) solar cells. Nanowire arrays can provide direct pathways for electron transport as well as sufficient surface area for sensitization. Progress in sensitized nanowire-based solar cells will be presented, as well as some of the challenges for improving their efficiencies. Overcoming these challenges will require detailed fundamental understanding and control of charge transfer in the materials and interfaces within the cells.

See more of this Session: Semiconductor Nanowires and Nanotubes for Next Generation Photovoltaics - From Fundamentals to Devices
See more of this Group/Topical: Topical 5: Nanomaterials for Energy Applications

(609d) Photoelectrochemical Complexes for Solar Energy Conversion That Dynamically, Reversibly and Autonomously Self-Assemble

Ardemis A. Boghossian[1], Jong Hyun Choi[2], Moon-Ho Ham[1], Esther S. Jeng[3], Rachel A. Graff[4], Daniel A. Heller[1], Alice C. Chang[3], Aidas Mattis[5], Timothy H. Bayburt[6], Yelena V. Grinkova[6], Adam Scott Zeiger[7], Krystyn J. Van Vliet[8], Erik K. Hobbie[9], Stephen G. Sligar[6], Colin A. Wraight[6] and Michael Strano[3], (1)Department of Chemical Engineering, Massachusetts Institute of Technology, Cambridge, MA, (2)Mechanical Engineering, Purdue University, West Lafayette, IN, (3)Chemical Engineering, Massachusetts Institute of Technology, Cambridge, MA, (4)Chemical and Biomolecular Engineering, University of Illinois at Urbana-Champaign, Urbana, IL, (5)Department of Biochemistry, University of Illinois at Urbana-Champaign, Urbana, IL, (6)Biochemistry, University of Illinois Urbana-Champaign, Urbana, IL, (7)Materials Science and Engineering, Massachusetts Institute of Technology, Cambridge, MA, (8)Materials Science and Engineering, MIT, Cambridge, MA, (9)Polymers Division, National Institute of Standards and Technology, Gaithersburg, MD

Photoelectrochemical complexes for solar energy conversion that dynamically, reversibly and autonomously self-assemble

Naturally occurring photosynthetic systems use elaborate pathways of self-repair to limit the impact of photo-damage. Here, we demonstrate a complex consisting of an aqueous solution containing photosynthetic reaction centers (RCs), membrane scaffold proteins (MSPs), phospholipids, and single-walled carbon nanotubes (SWCNTs) solubilized with the surfactant sodium cholate (SC) that mimics this process.(1) The components reversibly self-assemble into a highly ordered structure upon dialysis of the surfactant. The resulting assembled structure is photoelectrochemically active and consists of 4-nm-thick lipid bilayer disks (nanodisks, NDs) arranged parallel to the surface of the SWCNT with the RC housed within the bilayer such that its hole injecting site faces the nanotube surface. The structure can be assembled and disassembled autonomously with the addition or removal of surfactant. We model the kinetic and thermodynamic forces that drive the dynamics of this reversible self-assembly process.(2) The assembly is monitored using spectrofluorimetry during dialysis and subsequent surfactant addition and used to fit a kinetic model to determine the forward and reverse rate constants of ND and ND-SWCNT formation. The calculated ND and ND-SWCNT forward rate constants are 79 mM^{-1} s^{-1} and 5.4×10^2 mM^{-1} s^{-1}, respectively, and the reverse rate constants are negligible over the dialysis time scale. We find that the reaction is not diffusion-controlled since the ND-SWCNT reaction, which consists of entities with smaller diffusion coefficients, has a larger reaction rate constant. Using these rate parameters, we were able to develop a kinetic phase diagram for the formation of ND-SWCNT complexes, which indicates an optimal dialysis rate of approximately 8×10^{-4} s^{-1}.

The assembly is thermodynamically metastable and can only transition reversibly if the rate of surfactant removal exceeds a threshold value. Only in the assembled state do the complexes exhibit photoelectrochemical activity. We experimentally demonstrate a regeneration cycle that uses surfactant to switch between assembled and disassembled states, resulting in an increased photoconversion efficiency of more than 300% over 168 hours and an indefinite extension of the system lifetime. The system can disassemble and reassemble over an indefinite number of cycles. We also fit the model to these cyclic ND-SWCNT assembly and disassembly experiments and hence mimic the thermodynamic forces used in regeneration processes detailed previously. Such forces may form the basis of both synthetic and natural photoelectrochemical complexes capable of dynamic component replacement and repair.

1. Ham MH, Choi JH, Boghossian AA, Jeng ES, Graff R.A, Heller DA, Chang AC, Mattis A, Bayburt TH, Grinkova YV, Zeiger AS, Van Vliet KJ, Hobbie EK, Sligar SG, Wraight CA and **Strano MS**: Photoelectrochemical complexes for solar energy conversion that chemically and autonomously regenerate. NATURE CHEMISTRY, 2 (2010) 929-936.

2. Boghossian AA, Choi JH, Ham MH and **Strano MS**: Dynamical and Reversible Self-Assembly of Photoelectrochemical Complexes Based on Lipid Bilayer Disks, Photosynthetic Reaction Centers, and Single-Walled Carbon Nanotubes. Langmuir, 27 (2011) 1599-160

See more of this Session: Semiconductor Nanowires and Nanotubes for Next Generation Photovoltaics From Fundamentals to Devices
See more of this Group/Topical: Topical 5: Nanomaterials for Energy Applications

(609e) A Fundamental Investigation of Electron Transport and Recombination Characteristics of Titania Nanowire/Nanoparticle Hybrid Structures

Venkat Kalyan Vendra[1], Delaina A. Amos[1], Mahendra K. Sunkara[1] and Thad Druffel[2], (1)Chemical Engineering, University of Louisville, Louisville, KY, (2)Conn Center for Renewable Energy Research, Louisville

The electron transport and recombination of long titania nanowires and titania nanoparticle/nanowire composites are investigated. Understanding charge transport and recombination in 1D nanostructures of titania is critical for improving the performance of DSSC employing these 1D nanostructures as photoanodes in dye sensitized solar cells. 1D titania nanostructures are expected to have short transport times and high electron lifetimes. However, most of the 1D titania nanostructures have not shown these desirable properties. For instance, the electron diffusion coefficients and transport time constants in titania nanotubes were found to be very similar to that of titania nanoparticles.[1] But, the recombination time constants for titania nanotubes were 10 times higher than then nanoparticle based DSSCs. In contrast, DSSCs fabricated using titania nanowires showed both shorter transport time scales and higher electron lifetimes.[2] In a seminal paper, Richter and Schmuttenmaer showed that the lower the electron mobility in TiO2 nanotubes was due to the existence of exciton like trap states formed from Ti^{+3} donor states resulting from oxygen deficiencies, fluoride impurities and excess titanium introduced during synthesis.[3] Although thermal annealing is expected to remove the oxygen deficiencies, the structural integrity of the nanotubes would be compromised and hence a new synthesis technique is needed to prepare defect free TiO_2 nanotubes for enhanced DSSC performance.

Several approaches such as electrochemical anodization, solvothermal techniques, hydrothermal methods, thermal evaporation and template based approaches have been widely explored. However, most the methods produce short titania nanotubes (~ 2 μm in length) which is undesirable considering the low dye loading and slow electron transport in these architectures. In addition, all these synthesis techniques discussed above are very time consuming with the reaction time scales ranging from several hours to even days. To our knowledge there have not been any studies which have focused on the fast synthesis of single crystalline titania nanowires. Here we present a novel, scalable approach for making single crystalline titania nanowires arrays for dye sensitized solar cells.[4] Electron transport and recombination in these single crystalline nanowires and in single crystalline titania nanowires coated with titania nanoparticles are discussed.

[1] Zhu, K., Neale, N.R., Miedaner, A., Frank, A.J. Enhanced charge collection efficiencies and light scattering in dye sensitized solar cells using oriented TiO2 nanotubes. Nanoletters 2007, 7, 69-74.

[2] Tetreault, N., Horvath, E., Moehl, T., Brillet, J., Smajda, R., Bungener, S., Cai, N., Peng, W., Zakeeruddin, S.M., Forro, L., Magrez, A., Gratzel, M. High-Efficiency Solid-State Dye- Sensitized Solar Cells: Fast Charge Extraction through Self-Assembled 3D Fibrous Network of Crystalline TiO2 Nanowires. ACS Nano 2010, 4, 7644-7650.

[3] Richter, C., Schmuttenmaer, C.A., Exciton like trap states limit electron mobility in TiO2 nanotubes, Nature Nanotechnology 2010, 5, 769–772

[4] Kumar, V., Kim, J.H. Jasinski, J.B., Clark, E.L., Sunkara, M. Alkali metal assisted gas phase bulk production of titania nanowires. Journal of Crystal Growth and Design 2011, Article Accepted.

See more of this Session: Semiconductor Nanowires and Nanotubes for Next Generation Photovoltaics - From Fundamentals to Devices
See more of this Group/Topical: Topical 5: Nanomaterials for Energy Applications

(609f) Vertically Aligned Silicon Radial p-n Junction Micropillar Array Solar Cells

Kane J. Miller, Chemical Engineering, University of Louisville, Louisville, KY, Kevin M. Walsh, Department of Electrical and Computer Engineering, University of Lousiville, Lousiville, KY and Xiao-an Fu, Chemical Engineering, University of Lousiville, Lousiville, KY

Solar cell applications using silicon have been extensively studied in the past. The theoretical limit of the efficiency of single-crystalline silicon solar cells is well known. Several methods for optimizing silicon solar cells and obtaining the maximum possible efficiency exist in the literature. Among these methods are changes in the geometry and configuration of the solar cells, such as introducing radial p-n junction solar cells.

Micro- and nanoscale radial p-n junction solar cells, in theory, result in increased solar cell efficiency compared to planar p-n junctions by reducing carrier collection length. The geometry of arrays of these radial p-n junctions also introduces internal reflection and light trapping effects, ultimately improving the photon absorption characteristics.

We have fabricated a first-pass vertically-aligned micropillar p-n junction solar cell array with a micropillar diameter of 2 μm, length of approximately 10 μm, and distance of 2 μm between adjacent pillars. These arrays were fabricated using conventional photolithography and deep reactive ion etching on a (100) n-type silicon wafer, followed by boron diffusion using a spin-on dopant. Initial measurements demonstrated an open-circuit voltage of 0.36 V, short-circuit current density of 24 mA/ cm^2, a fill factor of 0.6, and efficiency of 5.3%.

Nanopillar p-n junction solar cell arrays of varying pillar diameters and lengths, fabricated using the combination of electron beam lithography and deep reactive ion etching, will be discussed. The test results will be compared to planar silicon solar cells and to theoretical calculations. With various techniques, we expect to increase the efficiencies of our solar cells beyond 10%.

See more of this Session: Semiconductor Nanowires and Nanotubes for Next Generation Photovoltaics - From Fundamentals to Devices
See more of this Group/Topical: Topical 5: Nanomaterials for Energy Applications

AUTHOR INDEX

AUTHOR INDEX

AUTHOR INDEX